TABLE OF CONTE

HomeTech Remodeling and Renovation Cost Estimator

Compiled and Edited by Henry Reynolds

Field Manual
1995
30th Annual Edition

Henry Reynolds
Editor

Ruth Miles Bruns
Publications Director

ISBN: 1-882379-14-4

HomeTech Information Systems, Inc.
5161 River Road, Bethesda, Maryland 20816
1-800-638-8292

ABBREVIATIONS

THE FOLLOWING ABBREVIATIONS ARE USED IN THIS MANUAL:

CF	PER CUBIC FOOT (ONE FOOT WIDE, ONE FOOT HIGH AND ONE FOOT LONG)
LF	PER LINEAL FOOT (PER RUNNING FOOT)
SF	PER SQUARE FOOT OF AREA
EA	EACH UNIT (A WINDOW UNIT INCLUDES WINDOW, FRAME, TRIM AND LABOR TO INSTALL)
EA PLUS SF	ADD THE AMOUNT OPPOSITE "EA" (EACH) TO THE TOTAL OBTAINED BY MULTIPLYING THE SQUARE FOOTAGE OF AREA BY THE AMOUNT OPPOSITE "SF"
UI	UNITED INCHES

TO CONVERT TO GROSS PROFITS OTHER THAN THE 33-1/3% RATE SHOWN IN THIS MANUAL, USE THE TABLE BELOW:

TO OBTAIN % OF GROSS PROFIT	MULTIPLY BOOK SALES PRICE BY
50%	1.34
45%	1.22
40%	1.12
35%	1.03
33-1/3%	- -
30%	.96
25%	.89
20%	.84

Preface to 30th

This estimating manual has been prepared for the use of contractors, estimators, supervisors, architects, insurance adjusters, urban renewal specialists, apartment house managers and all other professionals who need up-to-date cost information on which to base estimates and sales proposals.

Job costs include laying out the work, sweeping up and piling debris outside the building, but not removal of debris from premises unless so indicated.

A 33-1/3% gross profit is included in each sales price, obtained by adding 50% of the total labor and materials costs for each item. Thus, 2/3 of the sales price is cost, 1/3 is gross profit.

MATERIALS COSTS

The materials costs in this volume are based on current costs in a metropolitan area including applicable state, provincial and federal sales taxes.

The cost of materials includes everything that goes into the job. For example, drywall material would include the board, drywall nails, corner beads, tape, joint compound, even sandpaper. A normal waste factor is also included.

LABOR COSTS

The labor costs include:
HOURLY WAGES
WORKERS' COMP. INSURANCE
SOCIAL SECURITY TAXES
STATE UNEMPLOYMENT TAX
MINOR DAILY CLEANUP
COFFEE BREAKS
SET UP AND LAYOUT TIME
MINOR MATERIALS PICKUP

Also included are the normal inefficiencies of the remodeling business: time spent in discussions with customers on the job, placing and protecting delivered materials, assisting subcontractors, waiting for customer selections or official building inspections, etc.

Paid holidays, vacation and medical benefits are not included in the labor costs. If your employees receive these fringe benefits, you should calculate what these are as an hourly figure for 2,000 hours per year and add that to the labor cost shown in the manual.

SUBCONTRACTOR COSTS

The costs for plumbing, heating, electrical, built-up roofing, machine excavation, asphalt paving, ceramic tile and blown-in insulation are the amounts subcontractors in the trades would charge a general contractor.

The amounts in the "PRICE" columns throughout the volume are the amounts a general contractor would charge a homeowner or customer for the work. The costs shown, therefore, include the subcontractor's own profit on the job.

While many contractors do not charge a full markup for work done by subcontractors, it is our view that a remodeling contractor cannot make enough gross profit on a job to cover overhead unless everything, including subcontractors, is marked up at least 50%.

Annual Edition

The contractor has spent a good deal of time finding subcontractors who can do a job properly, on time, at a fair price. The contractor is responsible to the customer for quality of work if any problem arises later on. Should the subcontractor be unable or unwilling to correct the problem, the remodeler must do it at his own expense.

OVERHEAD ITEMS

Overhead items must be paid for out of the gross profit which is included in the amount shown in the "PRICE" column.

Overhead items include the following:
SALES COMMISSIONS
OFFICE SALARIES
JOB SUPERVISION
TRUCK EXPENSE
TELEPHONE
ADVERTISING AND MARKETING
OFFICE RENTAL
OFFICE SUPPLIES
LEGAL FEES
ACCOUNTANT FEES
LEASED OFFICE EQUIPMENT
TOOLS AND EQUIPMENT
INTEREST ON LOANS
BAD DEBTS
INSURANCE AND BONDS

JUDGEMENT FACTORS

Remodeling and renovation work is the most difficult to estimate in residential construction. There are more hidden problems resulting in additional costs than normally occur in new home construction. Care should be taken to determine whether structural defects exist inside closed floors, walls and ceilings before submitting a final bid.

Estimators must use judgement in working with this manual. Conditions under which jobs are performed in the building trades are variable, but normal, sunshiny job conditions and easy access to the job have been assumed in estimating the cost of each item. Important factors governing costs and bids received are the season of the year, general business conditions in the area, and the amount of work subcontractors have ahead when bidding your work.

Every effort is made to provide up to date, accurate cost information. Building costs throughout the United States and Canada are monitored, but cost changes occur frequently and are not always consistent with general economic conditions in an area. **No warranty or guarantee is made as to the correctness or sufficiency of the information contained in this book. The editors and publishers assume no responsibility or liability in connection with its use.**

The editors wish to thank all who have generously assisted in preparation of this volume. We are especially grateful to David Hawbecker, Darrell Lewis and Peter Starbuck for the materials and information which they have supplied. Constructive criticism which will assist us in the preparation of future editions will be welcomed.

Henry Reynolds, Editor
Bethesda, Maryland
December 1, 1994

How to Use the Remodeling and

SPEEDY RECKONER SECTION

Using the Speedy Reckoner, you can estimate many standard jobs quickly and accurately. For example, look at Page 2 to see how a standard shell addition is estimated. First, the specifications are given, clearly and completely, from the foundation through the exterior finish. Next, on Page 3, the pricing information is shown.

The Shell Addition has a base price of $6381, similar to a trip charge including set-up costs and minimum charges that occur regardless of the size of the addition. In addition to the base price there is a per-square-foot price which is $27.47 on a one-story addition. If you are estimating a 200 square foot addition, you would multiply 200 x $27.47 and add the result, $5494, to the $6381 base price for a total price of $11,875.

You will find the prices in the Speedy Reckoner accurate so long as the specifications for the job you are estimating are the same as the specifications in the manual. For example, the Shell Addition on Page 2 has no door, and one window per 100 SF of living area. If you are estimating a shell addition with more windows and one door, you will need to add the price of the additional windows and a door to the price calculated from the manual.

When used properly, the Speedy Reckoner can produce accurate estimates in as little as five minutes. It is also useful for "ballpark" estimates, and can be invaluable to estimators who do not know how to break down a job into its individual components.

SECTION II (ESTIMATOR)

This section of the manual is broken down into the 25 categories listed in the Table of Contents and on the back cover. These categories are organized in the order that a job is built, from Plans and Permits on through to Clean-Up. Many estimators use these 25 categories as a checklist to be sure they have not forgotten anything. The list of categories on the back cover matches up with black tab markings on the pages of the manual, so that you can easily flip to the category you want in seconds.

Here is an example of how an individual item is set up.

Look at Page 126, at the Exterior Stud Wall. In the SPECIFICATION section is a clear description of what is included in the item. The unit of measure, "SF", stands for square feet of wall (other items may be "LF" for lineal foot, "CF" for cubic foot, "EA" for each item, etc.). The Price includes a 50% markup.

TOTAL PRICE: The Price column marks up the Job Cost Total by 50% for a 33-1/3% gross profit margin. If you want to charge a higher or lower markup than 50%, see the table on Page ii.

HomeTech recommends a minimum markup of 50% over cost to make a profit in remodeling and renovation.

Renovation Cost Estimator

GENERAL INFORMATION

This manual is the best estimating guide available in the industry, and many contractors use it without adjustment. In order to use it properly, you must check it against the realities of your operation and adjust the figures as needed. For example, if the manual shows that drywall costs 75¢ per square foot, and your drywall hanger charges 87¢ per square foot, you should change the figure in the manual to 87¢ to reflect the reality.

Unit cost estimating allows you to estimate standard work very quickly and accurately. But that is only about 90% of a good, detailed bid. You need to apply a judgement analysis to any estimate to reflect special conditions. That analysis should include the following:

PROJECT ANALYSIS: If there are unusual characteristics to the job, such as a tricky roof tie-in or difficult materials matching, increase the job costs.

JOB CONDITIONS: Difficult conditions such as poor access, little or no storage, high risk of theft, etc., should be reflected in increased job costs.

CUSTOMER ANALYSIS: Approximately one out of five customers will be unusually demanding, and can cost you all the profit on a job unless you recognize them for what they are and increase your estimate to allow you to satisfy them and still make a profit.

COMPANY CAPABILITY: Certain kinds of work may be easier or harder than usual for your workers, and your estimate should be adjusted to reflect that. If you will need to hire more people or pay overtime in order to do a particular job, your increased costs should be reflected in your estimate.

If you have specific questions about using this manual, please call us. Use our toll-free number, 1-800-638-8292.

SAMPLE ESTIMATE

REMODELING SPEEDY RECKONER ESTIMATE

Name	Telephone	Job Telephone
J.M. KEYNES	555-3210	SAME
Address	**Estimator**	**Date**
474 MAIN ST., CITY	TRH	1 NOV 1994
Job Address	**Description of Work**	
SAME	SHELL ADDITION, BATHROOM, ELECTRICAL OUTLETS	

BASIC JOB, EXTRAS AND ALLOWANCES

Item	Unit	Unit Quantity	ADD		DEDUCT	
			Unit Price	PRICE	Unit Price	PRICE
BASIC PRICE	EACH	---	6381.	6,381		---
SQUARE FOOT AREA	SF	290	25.26	7,325		---
TEAR OUT MASONRY WALL	LF	3	96.	288		
INTERIOR STUD WALL 2x4, 16" O.C.	LF	10	16.56	166		
SUBSTITUTE BEVELED SIDING	SF	432	.86	372		
IN LIEU OF VINYL SIDING						
REAR DOOR 2-8x6-8	EA	1	612.	612		
BATH - W.C., LAV., TUB	TOTAL			3,210		
ELECTRICAL OUTLETS	EA	12	57.	684		
SWITCH & LIGHT	EA	1	111.	111		
INTER. BIRCH DOOR 2-6x6-8	EA	1	198.	198		
			Totals	19,347		
			Less Deductions	—		
			SALES PRICE	19,347		

HomeTech Form 136

HomeTech
Remodeling and Renovation
Cost Estimator

Section I
Speedy Reckoner

How to Use the Speedy Reckoner

The figures used in the Speedy Reckoner are the prices a contractor would charge a customer for the work. Prices include the contractor's 1/3 gross profit (50% markup over direct job costs).

After the basic price has been entered on your Speedy Reckoner estimate, there may be additional items that are not included in the basic job specifications. There may also be items in the basic price specifications that are to be eliminated in the particular job being estimated. Enter these items on your estimate sheet in the "ADD" or "DEDUCT" columns (see the sample Speedy Reckoner Estimate on Page -viii-). If the items are not under "Extras, Allowances, Alternate Specifications" in the Speedy Reckoner, look in Section II (starting on Page 74) and use the figures shown in the "PRICE" columns as the basis for computing the additions and deductions.

To Add Items Not Included In Basic Job:
Compute the price based on the actual quantity of the work to be done and set that amount down in the "ADD" column.

To Eliminate Items Included In Basic Job:
If an item specified in a basic job is not to be included in the job being estimated, you may eliminate it on your Speedy Reckoner Estimate as follows: compute what the total price of the item would have been if it had been included as specified and set that amount down in the "DEDUCT" column.

To Substitute One Item For Another In Basic Job:
When a different item is to be substituted for the item specified, refer to the Shell Addition Extras and Allowances on Pages 14 to 16.

For example, the Sample Shell Addition shown on Page -viii- calls for beveled wood siding on the sidewalls instead of the vinyl siding shown in the Shell Addition Specifications on Page 2. On Page 16, select the item that substitutes 6" beveled siding, and set down that price in the "ADD" column. (If you were substituting a less expensive siding, such as hardboard lap siding, you would set down the price from Page 16 in the "DEDUCT" column of your estimate.

SHELL ADDITION — ON BLOCK FOUNDATION WALL

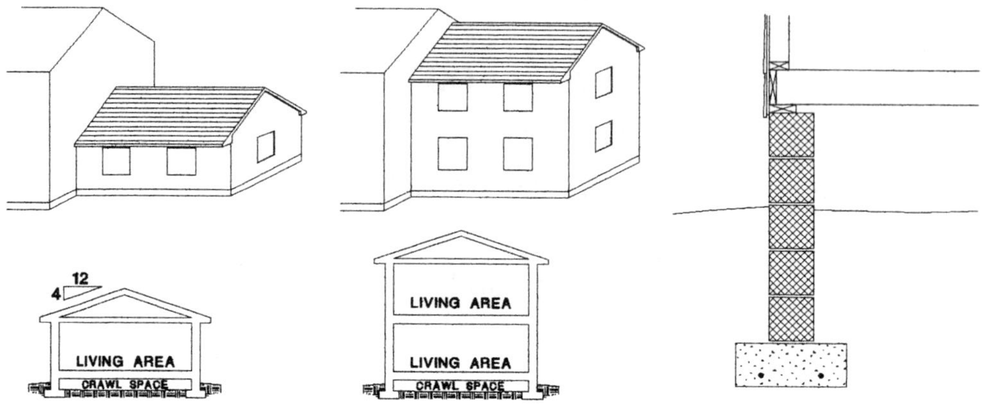

SPECIFICATIONS

PLANS AND PERMIT	PREPARE PLANS AND OBTAIN PERMIT, PERMIT FEE COST **NOT** INCLUDED
CONCRETE AND MASONRY	EXCAVATE BY HAND TO 36" BELOW GRADE, FORM AND POUR CONTINUOUS CONCRETE FOOTINGS WITH REBAR BLOCK FOUNDATION WALL ON FOOTINGS TO 18" ABOVE GRADE WITH 1/2" ANCHOR BOLTS 8'-0" O.C. PARGE, DAMPPROOF, BACKFILL AND ROUGH GRADE AS REQUIRED
FLOOR FRAMING	FLOOR JOISTS AS REQUIRED, 16" O.C. 3/4" T & G FIR PLYWOOD SUBFLOOR, GLUED AND NAILED
WALL FRAMING	2" X 4" STUDDING AND PLATES, 16" O.C. 1/2" CDX PLYWOOD OR FIBERBOARD SHEATHING WITH CORNER BRACING 1" X 2" FURRING, 16" O.C., ON EXISTING HOUSE WALL
ROOF FRAMING	SHED OR GABLE TYPE ROOF, 12" OVERHANG, **NO** OVERLAY OF EXISTING HOUSE ROOF RAFTERS AND CEILING JOISTS AS REQUIRED, 16" O.C. 1/2" CDX PLYWOOD ROOF SHEATHING
ROOFING, GUTTERS, FLASHING	#215 FIBERGLASS STRIP SHINGLES OVER #15 FELT PAPER ALUMINUM GUTTER AND DOWNSPOUT AS REQUIRED ALUMINUM FLASHING AT EXISTING HOUSE
EXTERIOR TRIM	TIGHT KNOT PINE OR FIR FASCIA AND SOFFIT TIGHT KNOT PINE OR FIR RAKE AND RAKE MOULDING
SIDING	8" VINYL SIDING (INSULATED)
WINDOW	**ONE** WOOD DOUBLE HUNG 2-8 X 4-6 WINDOW, DOUBLE GLAZED, FOR EACH 100 SF OF LIVING AREA
CLEAN-UP	REMOVE RUBBISH FROM PREMISES, DUMPING FEE **NOT** INCLUDED
	NO PAINTING

SHELL ADDITION — ON BLOCK FOUNDATION WALL

SPECIFICATIONS		UNIT	PRICE	LOCAL AREA MODIFICATION %	LOCAL AREA MODIFICATION PRICE
SHELL ADDITION	ONE STORY	EA	6381.00		
		PLUS			
	SF = SF LIVING AREA	SF	27.47		
	TWO STORY	EA	7785.00		
		PLUS			
	SF = SF LIVING AREA 2 FLOORS	SF	22.91		

EXTRAS AND ALLOWANCES

SPECIFICATIONS		UNIT	PRICE	LOCAL AREA MODIFICATION %	LOCAL AREA MODIFICATION PRICE
DIFFERENT FOOTING DEPTH	***Depth of Bottom of Footing Below Grade***				
	12" **DEDUCT**	LF	20.94		
	24" **DEDUCT**	LF	10.46		
	48" **ADD**	LF	10.43		
	LF = FOOTINGS				
CONCRETE FOUNDATION WALL	SUBSTITUTE CONCRETE FOUNDATION WALL IN PLACE OF BLOCK				
	Depth of Bottom of Footing Below Grade				
	24" **ADD**	LF	13.40		
	36" **ADD**	LF	12.83		
	48" **ADD**	LF	12.29		
	LF = WALL				
FOOTING AND COLUMN	EXCAVATE, FORM, POUR 24" X 24" X 12" CONCRETE FOOTING AND INSTALL 3" STEEL COLUMN UNDER BEAM **ADD**	EA	156.00		
WOOD BEAM	INSTALL WOOD BEAM TO SUPPORT JOISTS				
	6" X 8" **ADD**	LF	11.63		
	6" X 10" **ADD**	LF	17.94		
WINDOWS	ADD OR OMIT ONE WOOD DOUBLE HUNG WINDOW, 2-8 X 4-6, DOUBLE GLAZED **ADD** OR **DEDUCT**	EA	378.00		
	SUBSTITUTE DOUBLE GLAZED 4-0 X 4-0 ALUMINUM SLIDING WINDOW **DEDUCT**	EA	105.00		
	NOTE: FOR ADDITIONAL SHELL ADDITION EXTRAS AND ALLOWANCES, TURN TO PAGE 14				

SHELL ADDITION — SLAB ON GRADE

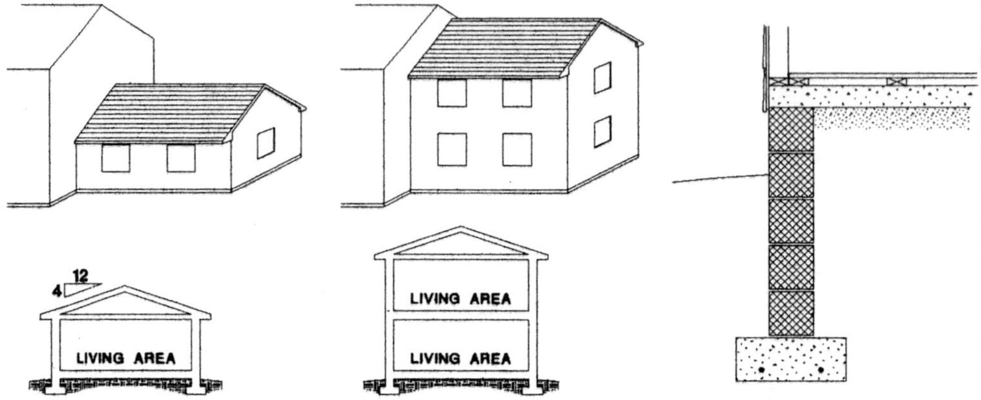

SPECIFICATIONS

PLANS AND PERMIT	PREPARE PLANS AND OBTAIN PERMIT, PERMIT FEE COST **NOT** INCLUDED
CONCRETE AND MASONRY	EXCAVATE BY HAND TO 36" BELOW GRADE, FORM AND POUR CONTINUOUS CONCRETE FOOTINGS WITH REBAR BLOCK FOUNDATION WALL ON FOOTINGS TO 18" ABOVE GRADE WITH 1/2" ANCHOR BOLTS 8'-0" O.C. 4" THICK CONCRETE FLOOR SLAB OVER GRAVEL BASE PARGE, DAMPPROOF, BACKFILL AND ROUGH GRADE AS REQUIRED
FLOOR FRAMING	2" X 4" SLEEPERS, 16" O.C., AND 3/4" T & G FIR PLYWOOD SUBFLOOR, GLUED AND NAILED OVER CONCRETE SLAB *TWO STORY SHELL ADDITION:* SECOND FLOOR JOISTS, 16" O.C. AND 3/4" PLYWOOD SUBFLOOR, GLUED AND NAILED
WALL FRAMING	2" X 4" STUDDING AND PLATES, 16" O.C. 1/2" CDX PLYWOOD OR FIBERBOARD SHEATHING WITH CORNER BRACING 1" X 2" FURRING, 16" O.C., ON EXISTING HOUSE WALL
ROOF FRAMING	SHED OR GABLE TYPE ROOF, 12" OVERHANG, **NO** OVERLAY OF EXISTING HOUSE ROOF RAFTERS AND CEILING JOISTS AS REQUIRED, 16" O.C. 1/2" CDX PLYWOOD ROOF SHEATHING
ROOFING, GUTTERS, FLASHING	#215 FIBERGLASS STRIP SHINGLES OVER #15 FELT PAPER ALUMINUM GUTTER AND DOWNSPOUT AS REQUIRED ALUMINUM FLASHING AT EXISTING HOUSE
EXTERIOR TRIM	TIGHT KNOT PINE OR FIR FASCIA AND SOFFIT TIGHT KNOT PINE OR FIR RAKE AND RAKE MOULDING
SIDING	8" VINYL SIDING (INSULATED)
WINDOW	**ONE** WOOD DOUBLE HUNG 2-8 X 4-6 WINDOW, DOUBLE GLAZED, FOR EACH 100 SF OF LIVING AREA
CLEAN-UP	REMOVE RUBBISH FROM PREMISES, DUMPING FEE **NOT** INCLUDED
	NO PAINTING

SHELL ADDITION — SLAB ON GRADE

SPECIFICATIONS		UNIT	PRICE	LOCAL AREA MODIFICATION %	LOCAL AREA MODIFICATION PRICE
SHELL ADDITION	ONE STORY SF = SF LIVING AREA	EA PLUS SF	6381.00 24.92		
	TWO STORY SF = SF LIVING AREA 2 FLOORS	EA PLUS SF	7785.00 20.61		

EXTRAS AND ALLOWANCES

SPECIFICATIONS		UNIT	PRICE	LOCAL AREA MODIFICATION %	LOCAL AREA MODIFICATION PRICE
DIFFERENT FOOTING DEPTH	***Depth of Bottom of Footing Below Grade***				
	12" **DEDUCT**	LF	20.94		
	24" **DEDUCT**	LF	10.46		
	48" **ADD**	LF	10.43		
	LF = FOOTINGS				
CONCRETE FOUNDATION WALL	SUBSTITUTE CONCRETE FOUNDATION WALL IN PLACE OF BLOCK				
	Depth of Bottom of Footing Below Grade				
	24" **ADD**	LF	13.40		
	36" **ADD**	LF	12.83		
	48" **ADD**	LF	12.29		
	LF = WALL				
OMIT SLEEPERS	OMIT SLEEPERS AND 5/8" PLYWOOD SUBFLOOR **DEDUCT** SF = SF FLOOR AREA	SF	3.81		
WINDOWS	ADD OR OMIT ONE WOOD DOUBLE HUNG WINDOW, 2-8 X 4-6, DOUBLE GLAZED **ADD** OR **DEDUCT**	EA	378.00		
	SUBSTITUTE DOUBLE GLAZED 4-0 X 4-0 ALUMINUM SLIDING WINDOW **DEDUCT**	EA	105.00		

NOTE: FOR ADDITIONAL SHELL ADDITION EXTRAS AND ALLOWANCES, TURN TO PAGE 14

SHELL ADDITION — ON MONOLITHIC (SINGLE POUR) SLAB

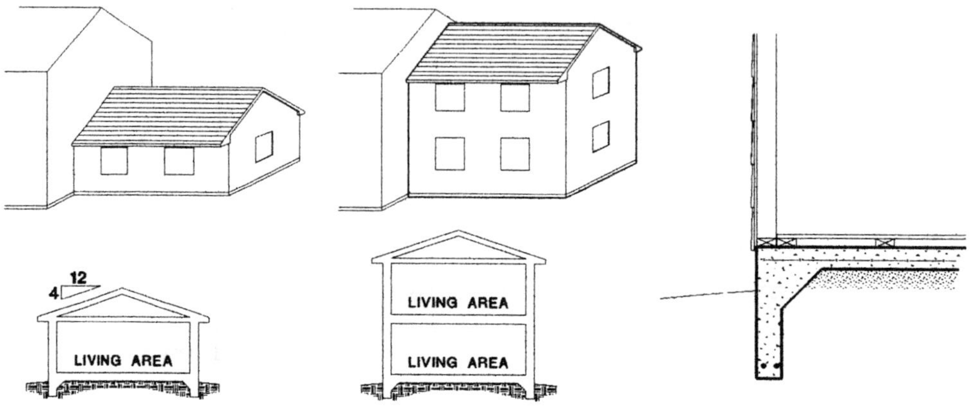

SPECIFICATIONS

PLANS AND PERMIT	PREPARE PLANS AND OBTAIN PERMIT, PERMIT FEE COST **NOT** INCLUDED
CONCRETE	EXCAVATE BY HAND FOR SINGLE POUR COMBINED FLOOR SLAB AND FOOTING FOUNDATIONS; SLAB THICKNESS 4" AND BOTTOM OF SLAB FOOTING 36" BELOW GRADE COVER AREA WITH 4" SAND OR GRAVEL AND PLASTIC VAPOR BARRIER PLACE 1/2" REINFORCING RODS IN FOOTING AREA AND WOVEN WIRE MESH IN SLAB & FOOTING AREA SET FORMS & POUR CONCRETE WITH 1/2" ANCHOR BOLTS 8'-0" O.C., SURFACE 8" ABOVE GRADE BACKFILL AND ROUGH GRADE AS REQUIRED
FLOOR FRAMING	2" X 4" SLEEPERS, 16" O.C., AND 3/4" T & G FIR PLYWOOD SUBFLOOR, GLUED AND NAILED OVER CONCRETE SLAB
WALL FRAMING	2" X 4" STUDDING AND PLATES, 16" O.C. 1/2" CDX PLYWOOD OR FIBERBOARD SHEATHING WITH CORNER BRACING 1" X 2" FURRING, 16" O.C., ON EXISTING HOUSE WALL
ROOF FRAMING	SHED OR GABLE TYPE ROOF, 12" OVERHANG, **NO** OVERLAY OF EXISTING HOUSE ROOF RAFTERS AND CEILING JOISTS AS REQUIRED, 16" O.C. 1/2" CDX PLYWOOD ROOF SHEATHING
ROOFING, GUTTERS, FLASHING	#215 FIBERGLASS STRIP SHINGLES OVER #15 FELT PAPER ALUMINUM GUTTER AND DOWNSPOUT AS REQUIRED ALUMINUM FLASHING AT EXISTING HOUSE
EXTERIOR TRIM	TIGHT KNOT PINE OR FIR FASCIA AND SOFFIT TIGHT KNOT PINE OR FIR RAKE AND RAKE MOULDING
SIDING	8" VINYL SIDING (INSULATED)
WINDOW	**ONE** WOOD DOUBLE HUNG 2-8 X 4-6 WINDOW, DOUBLE GLAZED, FOR EACH 100 SF OF LIVING AREA
CLEAN-UP	REMOVE RUBBISH FROM PREMISES, DUMPING FEE **NOT** INCLUDED
	NO PAINTING

SHELL ADDITION — ON MONOLITHIC (SINGLE POUR) SLAB

SPECIFICATIONS		UNIT	PRICE	LOCAL AREA MODIFICATION	
				%	PRICE
SHELL ADDITION	ONE STORY	EA	6381.00		
		PLUS			
	SF = SF LIVING AREA	SF	25.07		

EXTRAS AND ALLOWANCES

SPECIFICATIONS		UNIT	PRICE	LOCAL AREA MODIFICATION	
				%	PRICE
DIFFERENT FOOTING DEPTH	***Depth of Bottom of Footing Below Grade***				
	12" **DEDUCT**	SF	3.84		
	24" **DEDUCT**	SF	1.91		
	48" **ADD**	SF	1.98		
	SF = SHELL ADDITION FIRST FLOOR SF AREA				
OMIT SLEEPERS	OMIT SLEEPERS AND 5/8" PLYWOOD SUBFLOOR **DEDUCT**	SF	3.81		
	SF = SF FLOOR AREA				
WINDOWS	ADD OR OMIT ONE WOOD DOUBLE HUNG WINDOW, 2-8 X 4-6, DOUBLE GLAZED **ADD** OR **DEDUCT**	EA	378.00		
	SUBSTITUTE DOUBLE GLAZED 4-0 X 4-0 ALUMINUM SLIDING WINDOW **DEDUCT**	EA	105.00		

NOTE: FOR ADDITIONAL SHELL ADDITION EXTRAS AND ALLOWANCES, TURN TO PAGE 14

SHELL ADDITION — ON BRICK/BLOCK OR ROUND CONCRETE PIERS

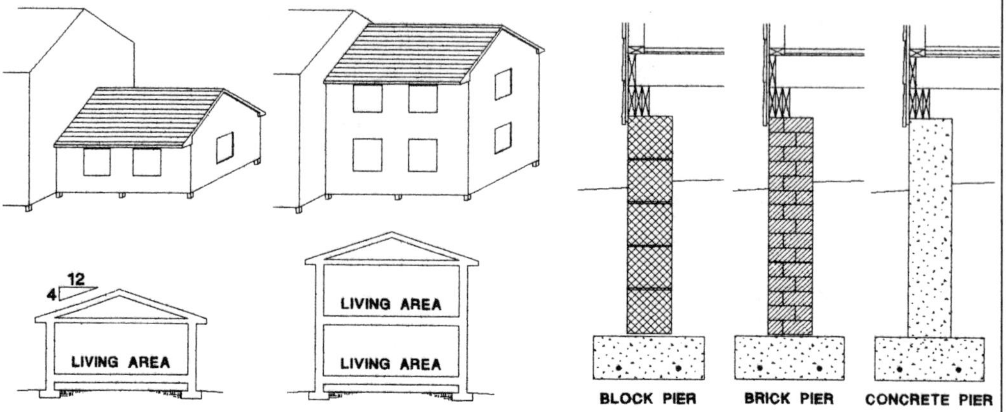

SPECIFICATIONS

PLANS AND PERMIT	PREPARE PLANS AND OBTAIN PERMIT, PERMIT FEE COST **NOT** INCLUDED
CONCRETE AND MASONRY	EXCAVATE BY HAND TO 36" BELOW GRADE, FORM AND POUR 12" X 24" X 24" CONCRETE PIER FOOTINGS BLOCK, BRICK OR CONCRETE IN SONOTUBE PIERS 8'-0" O.C. TO 18" ABOVE GRADE, WITH 1/2" ANCHOR BOLT IN EACH PIER
FLOOR FRAMING	2" X 8" SILL BEAM ON PIERS FASTENED TO ANCHOR BOLTS FLOOR JOISTS AS REQUIRED, 16" O.C. 3/4" T & G FIR PLYWOOD SUBFLOOR, GLUED AND NAILED
WALL FRAMING	2" X 4" STUDDING AND PLATES, 16" O.C. 1/2" CDX PLYWOOD OR FIBERBOARD SHEATHING WITH CORNER BRACING 1" X 2" FURRING, 16" O.C., ON EXISTING HOUSE WALL
ROOF FRAMING	SHED OR GABLE TYPE ROOF, 12" OVERHANG, **NO** OVERLAY OF EXISTING HOUSE ROOF RAFTERS AND CEILING JOISTS AS REQUIRED, 16" O.C. 1/2" CDX PLYWOOD ROOF SHEATHING
ROOFING, GUTTERS, FLASHING	#215 FIBERGLASS STRIP SHINGLES OVER #15 FELT PAPER ALUMINUM GUTTER AND DOWNSPOUT AS REQUIRED ALUMINUM FLASHING AT EXISTING HOUSE
EXTERIOR TRIM	TIGHT KNOT PINE OR FIR 6" FASCIA AND 12" SOFFIT TIGHT KNOT PINE OR FIR RAKE AND RAKE MOULDING
SIDING	8" VINYL SIDING (INSULATED)
WINDOW	**ONE** WOOD DOUBLE HUNG 2-8 X 4-6 WINDOW, DOUBLE GLAZED, FOR EACH 100 SF OF LIVING AREA
CLEAN-UP	REMOVE RUBBISH FROM PREMISES, DUMPING FEE **NOT** INCLUDED
	NO PAINTING

SHELL ADDITION — ON BRICK/BLOCK OR ROUND CONCRETE PIERS

SPECIFICATIONS		UNIT	PRICE	LOCAL AREA MODIFICATION %	LOCAL AREA MODIFICATION PRICE
SHELL ADDITION ON BRICK PIERS	ONE STORY	EA PLUS SF	5850.00 20.36		
	TWO STORY	EA PLUS SF	4875.00 22.43		
SHELL ADDITION ON BLOCK PIERS	ONE STORY	EA PLUS SF	5850.00 19.95		
	TWO STORY	EA PLUS SF	4875.00 22.22		
SHELL ADDITION ON ROUND CONCRETE PIERS	ONE STORY SF = SF LIVING AREA	EA PLUS SF	5850.00 19.89		
	TWO STORY SF = TOTAL SQUARE FOOT LIVING AREA, ONE OR TWO FLOORS	EA PLUS SF	4875.00 22.19		

EXTRAS AND ALLOWANCES

SPECIFICATIONS		UNIT	PRICE	LOCAL AREA MODIFICATION %	LOCAL AREA MODIFICATION PRICE
DIFFERENT FOOTING DEPTH	***Depth of Bottom of Footing Below Grade***				
	12" **DEDUCT**	SF	1.20		
	24" **DEDUCT**	SF	.90		
	48" **ADD**	SF	.90		
	SF = SHELL ADDITION FIRST FLOOR SF AREA				
FOOTING AND COLUMN	EXCAVATE, FORM, POUR 24" X 24" X 12" CONCRETE FOOTING & INSTALL 3" STEEL COLUMN UNDER BEAM **ADD**	EA	156.00		
WOOD BEAM	INSTALL WOOD BEAM TO SUPPORT JOISTS				
	6" X 8" **ADD**	LF	11.63		
	6" X 10" **ADD**	LF	17.94		
WINDOWS	ADD OR OMIT ONE WOOD DOUBLE HUNG WINDOW, 2-8 X 4-6, DOUBLE GLAZED **ADD** OR **DEDUCT**	EA	378.00		
	SUBSTITUTE DOUBLE GLAZED 4-0 X 4-0 ALUMINUM SLIDING WINDOW **DEDUCT**	EA	105.00		
	NOTE: FOR ADDITIONAL SHELL ADDITION EXTRAS AND ALLOWANCES, TURN TO PAGE 14				

SHELL ADDITION — OVER FULL BASEMENT

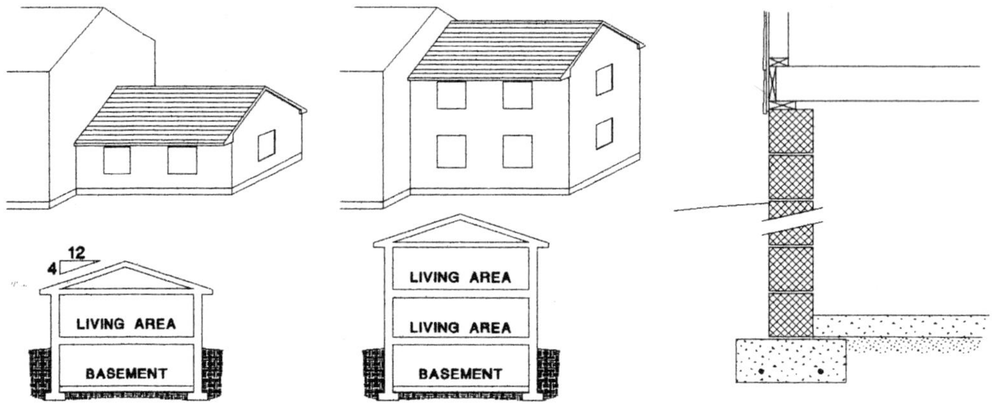

SPECIFICATIONS

PLANS AND PERMIT	PREPARE PLANS AND OBTAIN PERMIT, PERMIT FEE COST **NOT** INCLUDED
CONCRETE AND MASONRY	EXCAVATE FULL BASEMENT WITH FRONT END LOADER OR BACKHOE UNDER LIVING AREA, AND INSTALL CONTINUOUS CONCRETE FOOTINGS WITH REBAR BLOCK FOUNDATION WALL ON FOOTINGS TO 18" ABOVE GRADE WITH 1/2" ANCHOR BOLTS 8'-0" O.C. 4" CONCRETE BASEMENT FLOOR OVER GRAVEL BASE PARGE, DAMPPROOF, BACKFILL AND ROUGH GRADE AS REQUIRED
FLOOR FRAMING	FLOOR JOISTS AS REQUIRED, 16" O.C. 3/4" T & G FIR PLYWOOD SUBFLOOR, GLUED AND NAILED
WALL FRAMING	2" X 4" STUDDING AND PLATES, 16" O.C. 1/2" CDX PLYWOOD OR FIBERBOARD SHEATHING WITH CORNER BRACING 1" X 2" FURRING, 16" O.C., ON EXISTING HOUSE WALL
ROOF FRAMING	SHED OR GABLE TYPE ROOF, 12" OVERHANG, **NO** OVERLAY OF EXISTING HOUSE ROOF RAFTERS AND CEILING JOISTS AS REQUIRED, 16" O.C. 1/2" CDX PLYWOOD ROOF SHEATHING
ROOFING, GUTTERS, FLASHING	#215 FIBERGLASS STRIP SHINGLES OVER #15 FELT PAPER ALUMINUM GUTTER AND DOWNSPOUT AS REQUIRED ALUMINUM FLASHING AT EXISTING HOUSE
EXTERIOR TRIM	TIGHT KNOT PINE OR FIR FASCIA AND SOFFIT TIGHT KNOT PINE OR FIR RAKE AND RAKE MOULDING
SIDING	8" VINYL SIDING (INSULATED)
WINDOW	**ONE** WOOD DOUBLE HUNG 2-8 X 4-6 WINDOW, DOUBLE GLAZED, FOR EACH 100 SF OF LIVING AREA **ONE** WOOD FRAMED SINGLE GLAZED BASEMENT WINDOW FOR EACH 150 SF OF BASEMENT FLOOR AREA
CLEAN-UP	REMOVE RUBBISH FROM PREMISES, DUMPING FEE **NOT** INCLUDED
	NO PAINTING

SHELL ADDITION — OVER FULL BASEMENT

SPECIFICATIONS		UNIT	PRICE	LOCAL AREA MODIFICATION %	LOCAL AREA MODIFICATION PRICE
SHELL ADDITION	ONE STORY SF = SF LIVING AREA	EA PLUS SF	6381.00 43.80		
	TWO STORY SF = SF LIVING AREA 2 FLOORS	EA PLUS SF	7785.00 30.06		

EXTRAS AND ALLOWANCES

SPECIFICATIONS		UNIT	PRICE	LOCAL AREA MODIFICATION %	LOCAL AREA MODIFICATION PRICE
CONCRETE	SUBSTITUTE CONCRETE FOUNDATION WALL IN PLACE OF BLOCK **ADD** LF = WALL	LF	32.25		
DOORWAY TO EXISTING BASEMENT	BREAK THROUGH EXISTING HOUSE FOUNDATION WALL TO CONNECT WITH BASEMENT OF SHELL ADDITION AND TRIM OPENING WITH JAMBS AND 2-1/4" CASINGS				
	BLOCK WALL **ADD**	EA	180.00		
	CONCRETE WALL **ADD**	EA	279.00		
	EA = TOTAL COST EACH OPENING				
FOOTING AND COLUMN	EXCAVATE, FORM, POUR 24" X 24" X 12" CONCRETE FOOTING AND INSTALL 3" STEEL COLUMN UNDER BEAM **ADD**	EA	156.00		
WOOD BEAM	INSTALL WOOD BEAM TO SUPPORT JOISTS				
	6" X 8" **ADD**	LF	11.63		
	6" X 10" **ADD**	LF	17.94		
WINDOWS	ADD OR OMIT ONE WOOD DOUBLE HUNG WINDOW, 2-8 X 4-6, DOUBLE GLAZED **ADD** OR **DEDUCT**	EA	378.00		
	SUBSTITUTE DOUBLE GLAZED 4-0 X 4-0 ALUMINUM SLIDING WINDOW **DEDUCT**	EA	105.00		
	ADD OR OMIT ONE WOOD FRAMED BASEMENT WINDOW **ADD** OR **DEDUCT**	EA	127.50		

NOTE: FOR ADDITIONAL SHELL ADDITION EXTRAS AND ALLOWANCES, TURN TO PAGE 14

2nd STORY ADDITION (OVER EXISTING 1st FLOOR)

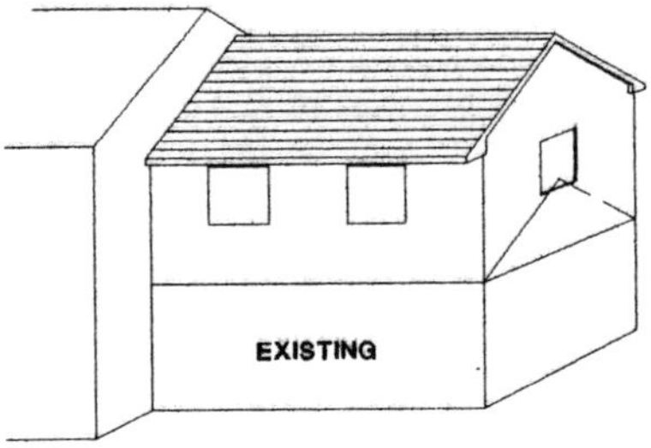

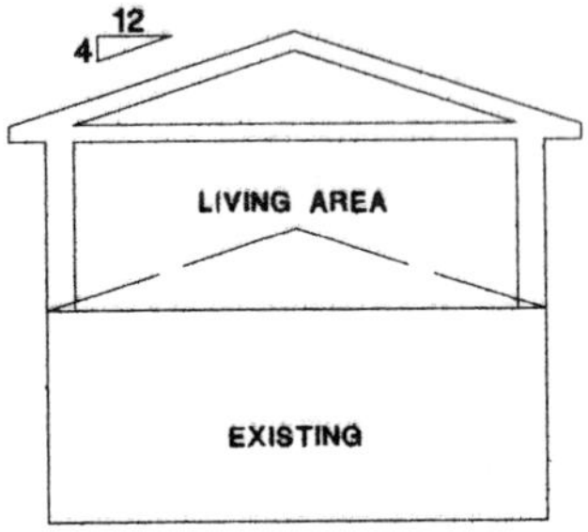

SPECIFICATIONS	
PLANS AND PERMIT	PREPARE PLANS AND OBTAIN PERMIT, PERMIT FEE COST **NOT** INCLUDED
TEAR-OUT	REMOVE EXISTING ROOF WHERE NECESSARY, INCLUDING OVERHANG AND EAVES REMOVE EXISTING FASCIAS AND SOFFITS
FLOOR FRAMING	INSTALL NEW FLOOR JOISTS FOR SECOND FLOOR, OR SISTER (DOUBLE UP) EXISTING JOISTS 3/4" T & G FIR PLYWOOD SUBFLOOR, GLUED AND NAILED
WALL FRAMING	2" X 4" STUDDING AND PLATES, 16" O.C. 1/2" CDX PLYWOOD OR FIBERBOARD SHEATHING WITH CORNER BRACING 1" X 2" FURRING, 16" O.C., ON EXISTING HOUSE WALL
ROOF FRAMING	SHED OR GABLE TYPE ROOF, 12" OVERHANG, **NO** OVERLAY OF EXISTING HOUSE ROOF RAFTERS AND CEILING JOISTS AS REQUIRED, 16" O.C. 1/2" CDX PLYWOOD ROOF SHEATHING
ROOFING, GUTTERS, FLASHING	#215 FIBERGLASS STRIP SHINGLES OVER #15 FELT PAPER ALUMINUM GUTTER AND DOWNSPOUT AS REQUIRED ALUMINUM FLASHING AT EXISTING HOUSE
EXTERIOR TRIM	TIGHT KNOT PINE OR FIR FASCIA AND SOFFIT TIGHT KNOT PINE OR FIR RAKE AND RAKE MOULDING
SIDING	8" VINYL SIDING (INSULATED) ON 3 NEW WALLS AND ON EXISTING HOUSE WALLS BELOW 2nd STORY ADDITION
WINDOW	**ONE** WOOD DOUBLE HUNG 2-8 X 4-6 WINDOW, DOUBLE GLAZED, FOR EACH 100 SF OF LIVING AREA
CLEAN-UP	REMOVE RUBBISH FROM PREMISES, DUMPING FEE **NOT** INCLUDED
	NO PAINTING

2nd STORY ADDITION (OVER EXISTING 1st FLOOR)

SPECIFICATIONS		UNIT	PRICE	LOCAL AREA MODIFICATION %	LOCAL AREA MODIFICATION PRICE
SHELL ADDITION	SECOND STORY ADDITION OVER EXISTING FIRST FLOOR SF = SF LIVING AREA, 2nd FLOOR ONLY	EA PLUS SF	5850.00 29.85		

EXTRAS AND ALLOWANCES

SPECIFICATIONS		UNIT	PRICE	LOCAL AREA MODIFICATION %	LOCAL AREA MODIFICATION PRICE
OMIT JOISTS	OMIT INSTALLATION OF NEW JOISTS OR SISTER-ING EXISTING JOISTS				
	2" X 8" **DEDUCT**	SF	5.13		
	2" X 10" **DEDUCT**	SF	5.72		
	2" X 12" **DEDUCT**	SF	6.21		
SUBFLOOR	REMOVE EXISTING SUBFLOOR				
	1" X 6" OR 1" X 8" **ADD**	SF	.78		
	NAILED PLYWOOD **ADD**	SF	.68		
	SF = FLOOR				
	INSTALL 3/4" T & G FIR PLYWOOD SUBFLOOR **ADD** SF = FLOOR	SF	1.94		
SIDING	OMIT SIDING ON EXISTING HOUSE WALL BELOW 2nd STORY ADDITION **DEDUCT** SF = SF EXISTING HOUSE WALL	SF	4.13		
WINDOWS	ADD OR OMIT ONE WOOD DOUBLE HUNG WIN-DOW, 2-8 X 4-6, DOUBLE GLAZED **ADD** OR **DEDUCT**	EA	378.00		
	SUBSTITUTE DOUBLE GLAZED 4-0 X 4-0 ALUMINUM SLIDING WINDOW **DEDUCT**	EA	105.00		

NOTE: FOR ADDITIONAL SHELL ADDITION EXTRAS AND ALLOWANCES, TURN TO PAGE 14

SHELL ADDITION — EXTRAS AND ALLOWANCES

SPECIFICATIONS			UNIT	PRICE	LOCAL AREA MODIFICATION	
					%	PRICE
SUBSTITUTE GABLE/SHED ROOF SLOPE AND/OR COVERING	SF = SF FLOOR AREA OF TOP FLOOR OF ADDITION ON WHICH THE RAFTERS ARE SUPPORTED. INCLUDES ADDITIONAL GABLE END FRAMING AND SIDING.					
	Same Framing, Same Slope					
	PENNSYLVANIA SLATE	**ADD**	SF	8.43		
	16" CEDAR SHINGLES	**ADD**	SF	4.50		
	CONCRETE TILES	**ADD**	SF	2.31		
	6/12 Slope					
	ASPHALT/FIBERGLASS	**ADD**	SF	1.22		
	PENNSYLVANIA SLATE	**ADD**	SF	2.01		
	16" CEDAR SHINGLES	**ADD**	SF	1.62		
	CONCRETE TILES	**ADD**	SF	1.43		
	8/12 Slope					
	ASPHALT/FIBERGLASS	**ADD**	SF	1.68		
	PENNSYLVANIA SLATE	**ADD**	SF	2.88		
	16" CEDAR SHINGLES	**ADD**	SF	2.30		
	CONCRETE TILES	**ADD**	SF	2.03		
	10/12 Slope					
	ASPHALT/FIBERGLASS	**ADD**	SF	2.72		
	PENNSYLVANIA SLATE	**ADD**	SF	4.70		
	16" CEDAR SHINGLES	**ADD**	SF	3.74		
	CONCRETE TILES	**ADD**	SF	3.26		
	12/12 Slope					
	ASPHALT/FIBERGLASS	**ADD**	SF	3.71		
	PENNSYLVANIA SLATE	**ADD**	SF	6.50		
	16" CEDAR SHINGLES	**ADD**	SF	5.13		
	CONCRETE TILES	**ADD**	SF	4.46		
SUBSTITUTE HIP ROOF FRAMING & ROOF COVERING	SF = SF FLOOR AREA OF TOP FLOOR OF ADDITION ON WHICH THE RAFTERS ARE SUPPORTED. INCLUDES OMITTING GABLE END FRAMING AND SIDING					
	SAME SLOPE	**ADD**	SF	.26		
	6/12 SLOPE	**ADD**	SF	.36		
	8/12 SLOPE	**ADD**	SF	.42		
	10/12 SLOPE	**ADD**	SF	.51		
	12/12 SLOPE	**ADD**	SF	.60		
SUBSTITUTE FLAT ROOF FRAMING & ROOF COVERING	SF = SF FLOOR AREA OF TOP FLOOR OF ADDITION ON WHICH THE RAFTERS (JOISTS) ARE SUPPORTED. INCLUDES OMITTING GABLE END FRAMING AND SIDING					
	1/12 Slope					
	4-PLY BUILT UP	**ADD**	SF	2.61		
	SINGLE PLY MEMBRANE	**ADD**	SF	2.13		
	16 OZ. COPPER	**ADD**	SF	14.39		
	GALV. SHEET METAL	**ADD**	SF	3.33		

SHELL ADDITION — EXTRAS AND ALLOWANCES

SPECIFICATIONS		UNIT	PRICE	LOCAL AREA MODIFICATION %	LOCAL AREA MODIFICATION PRICE
OVERLAY EXISTING ROOF WITH FRAMING & ROOFING	ASPHALT/FIBERGLASS **ADD**	SF	12.33		
	PENNSYLVANIA SLATE **ADD**	SF	22.88		
	16" CEDAR SHINGLES **ADD**	SF	16.67		
	CONCRETE TILES **ADD**	SF	13.68		
	SF = SF AREA OF ROOF OVERLAY				
TEAR-OUT	REMOVE WOOD OR ALUMINUM SIDING FROM EXISTING HOUSE WALL **ADD**	SF	.51		
	SF = SF OF HOUSE WALL				
CONCRETE SLAB AND STEPS	EXCAVATE, FORM, POUR AND FINISH CONCRETE SLAB, ONE OR TWO STEPS TO ADDITION, SLAB 6'-0" X 4'-0"				
	Depth of Bottom of Footing Below Grade				
	12" **ADD**	EA	594.00		
	24" **ADD**	EA	687.00		
	36" **ADD**	EA	807.00		
	48" **ADD**	EA	930.00		
	EA = TOTAL AMOUNT				
2" X 6" STUDS	SUBSTITUTE 2" X 6" STUDS FOR SPECIFIED 2" X 4" **ADD**	LF	.57		
	LF = LENGTH OF WALL				
PARTITION	FRAME OUT PARTITION IN ROOM WITH STUDS AND PLATES				
	2" X 4" **ADD**	LF	16.56		
	2" X 6" **ADD**	LF	21.12		
OMIT FURRING	OMIT FURRING OF EXISTING HOUSE WALL **DEDUCT**	SF	.92		
	SF = EXISTING HOUSE WALL				
SUBSTITUTE OVERHANG AND SOFFIT	ADDITIONAL, OR LESS, ROOF OVERHANG AND SOFFIT, INCLUDING ADDITIONAL OR LESS ROOF FRAMING AND ROOF COVERING				
	6" **DEDUCT**	LF	5.25		
	18" **ADD**	LF	16.20		
	24" **ADD**	LF	21.68		
	LF = LF OF SOFFIT				

SHELL ADDITION — EXTRAS AND ALLOWANCES

SPECIFICATIONS			UNIT	PRICE	LOCAL AREA MODIFICATION	
					%	PRICE
SUBSTITUTE SIDING	6" BEVELED SIDING	**ADD**	SF	.86		
	BRICK VENEER	**ADD**	SF	6.83		
	STUCCO	**ADD**	SF	4.85		
	18" CEDAR SHINGLES	**ADD**	SF	.54		
	3/8" FIR TEXTURE 1-11	**DEDUCT**	SF	1.58		
	HARDBOARD LAP SIDING	**DEDUCT**	SF	1.20		
	SF = SF OF SIDING					
FRAME DOORWAY	REMOVE EXISTING WINDOW AND WALL UNDER AND FRAME OUT FOR DOORWAY FROM HOUSE TO ADDITION	**ADD**	EA	216.00		
	BREAK THROUGH EXISTING HOUSE WALL AND FRAME OUT FOR DOORWAY FROM HOUSE TO ADDITION					
	FRAME WALL	**ADD**	EA	327.00		
	MASONRY WALL	**ADD**	EA	417.00		

ADDITION — TRIM AND INTERIOR FINISHING

SPECIFICATIONS	
WINDOWS	TRIM WINDOWS (ONE FOR EACH 100 SQUARE FEET OF LIVING AREA) WITH 2-1/4" CASING, STOOLS, STOPS AND SASH FASTENER
HEATING	EXTEND HEAT AND RETURN DUCTS FROM EXISTING SERVICE
ELECTRIC	DUPLEX WALL OUTLETS TO CODE ON EXISTING SERVICE
INSULATION	3-1/2" FIBERGLASS BLANKET INSULATION ON ALL SIDEWALLS 6" FIBERGLASS BLANKET INSULATION IN CEILING
INTERIOR WALL COVERING	1/2" GYPSUM BOARD (SHEETROCK) ON NEW WALLS AND ON FURRING STRIPS OVER EXISTING HOUSE WALL, TAPED AND FINISHED (**NO** PAINTING)
CEILING COVERING	1/2" GYPSUM BOARD (SHEETROCK) TAPED AND FINISHED (**NO** PAINTING)
INTERIOR TRIM	3-1/2" BASE AND SHOE MOULDING ON ALL FOUR WALLS
FLOOR COVERING	2-1/4" SELECT OAK FLOORING, SANDED AND FINISHED
CLEAN-UP	REMOVE TRASH FROM PREMISES, DUMPING FEE **NOT** INCLUDED
	NO PAINTING

SPECIFICATIONS		UNIT	PRICE	LOCAL AREA MODIFICATION	
				%	PRICE
INTERIOR TRIM AND FINISHING	AS SPECIFIED	EA PLUS	1305.00		
	SF = TOTAL SQUARE FOOT LIVING AREA TO BE FINISHED	SF	20.00		

ADDITION — TRIM & FINISHING EXTRAS AND ALLOWANCES

SPECIFICATIONS		UNIT	PRICE	LOCAL AREA MODIFICATION %	LOCAL AREA MODIFICATION PRICE
INSTALL DOOR FROM ADDITION TO EXISTING HOUSE	IN EXISTING FRAMED OPENING, INSTALL 2-6 X 6-8 X 1-3/8" FLUSH HOLLOW CORE BIRCH DOOR, INCLUDING JAMBS, STOPS, TRIM, HARDWARE & PATCHING **ADD**	EA	282.00		
INSTALL DOOR FROM ADDITION TO EXTERIOR	IN EXISTING FRAMED OPENING, INSTALL DOOR TO OUTSIDE — 2-8 X 6-8 X 1-3/4" 3 PANEL, 4 LIGHT FIR DOOR **ADD**	EA	612.00		
	GLIDING PATIO DOOR, ALUMINUM FRAMED, WITH INSULATED GLASS AND INSECT SCREEN, 6-0 X 6-8 **ADD**	EA	735.00		
DOOR	HANG DOOR IN FRAMED OUT PARTITION, 2-6 X 6-8, BIRCH HOLLOW CORE, INCLUDING JAMBS, STOPS, CASINGS AND HARDWARE **ADD**	EA	237.00		
	BI-FOLD DOORS, BIRCH FLUSH, INCLUDING ALL TRIM AND HARDWARE				
	2 DOORS 3-0 X 6-8 **ADD**	SET	216.00		
	6-0 X 6-8 **ADD**	SET	264.00		
	4 DOORS 4-0 X 6-8 **ADD**	SET	255.00		
	6-0 X 6-8 **ADD**	SET	285.00		
WINDOW	TRIM ADDITIONAL WINDOW WITH 2-1/4" CASINGS AND APRON, STOOL AND STOPS **ADD**	EA	75.00		
HEATING	OMIT EXTENDING HEAT TO ADDITION **DEDUCT**	SF	2.85		
	SF = SF LIVING AREA OF ADDITION				
INSULATION	INSULATE FLOOR JOISTS WITH FIBERGLASS BLANKET INSULATION				
	6" **ADD**	SF	.92		
	9" **ADD**	SF	1.19		
	SF = FLOOR JOISTS				

ADDITION — TRIM & FINISHING EXTRAS AND ALLOWANCES

SPECIFICATIONS		UNIT	PRICE	LOCAL AREA MODIFICATION	
				%	PRICE
WALL COVERING	OMIT DRYWALL ON EXISTING HOUSE WALL **DEDUCT** SF = EXISTING HOUSE WALL	SF	1.31		
INTERIOR PARTITION	TRIM AND FINISH STUD PARTITION TWO SIDES OF WALL • DUPLEX OUTLETS TO CODE • 1/2" DRYWALL, TAPED AND FINISHED • 3-1/2" BASE AND SHOE MOULD **ADD** LF = LENGTH OF 8-FOOT HIGH PARTITION	LF	46.50		
CLOSET TRIM	TRIM FRAMED CLOSET WITH BASE AND SHOE, HOOKSTRIP, SHELF, CLOTHES POLE AND CLOTHES POLE SOCKET	LF	23.67		
CEILING MOULDING	CEILING MOULDING AT WALL INTERSECTIONS 3/4" **ADD**	LF	1.76		
	1-5/8" **ADD**	LF	2.28		
	3-5/8" **ADD**	LF	3.78		
SUBSTITUTE FLOOR COVERING	SUBSTITUTE CARPET AND PAD OVER SUBFLOOR CARPET @ $15/YD **DEDUCT**	SF	7.11		
	@ $25/YD **DEDUCT**	SF	4.85		
	SUBSTITUTE VINYL TILE OVER SUBFLOOR **DEDUCT**	SF	5.54		

3 AND 4 FIXTURE BATHROOMS

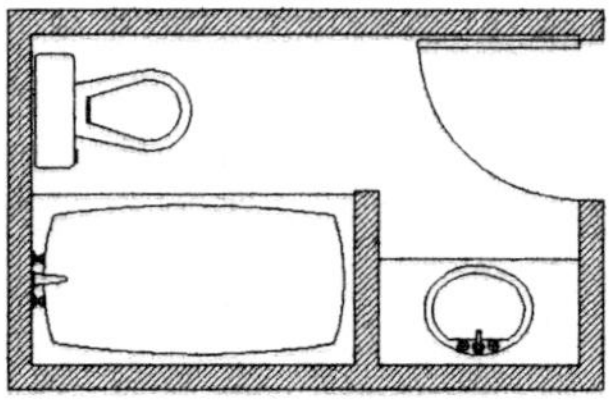

NO. 1 - 44 SF

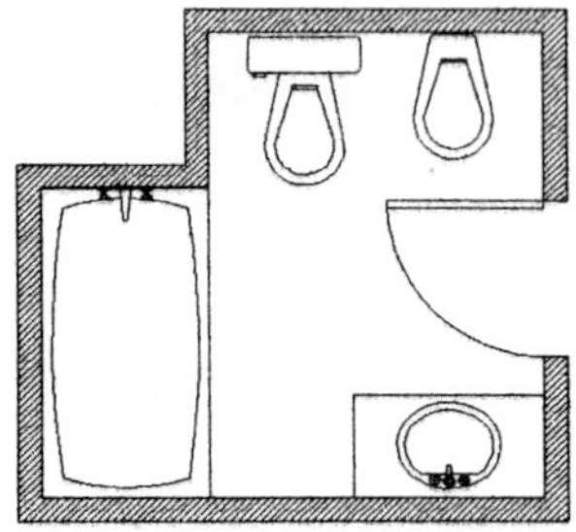

NO. 2 - 53 SF

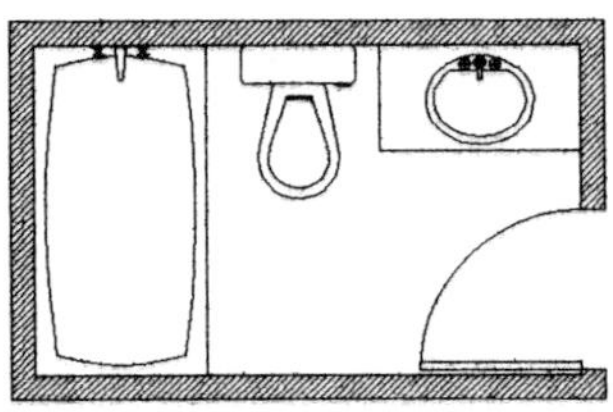

NO. 3 - 43 SF

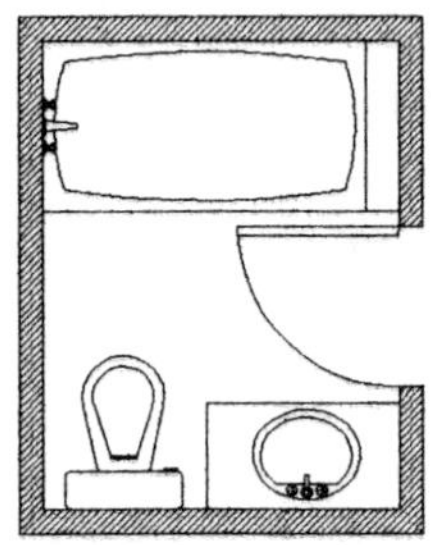

NO. 4 - 40 SF

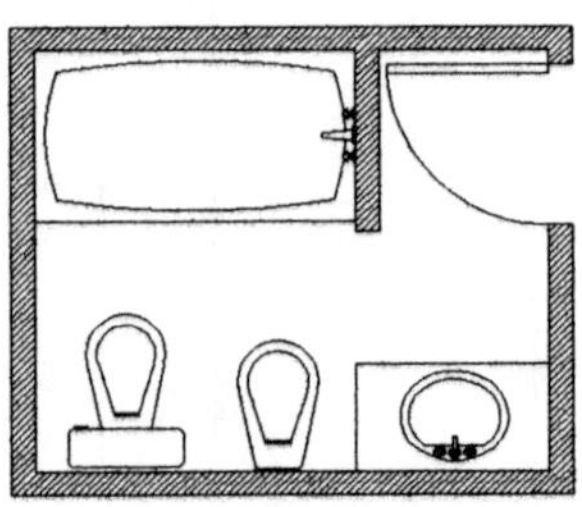

NO. 5 - 52 SF

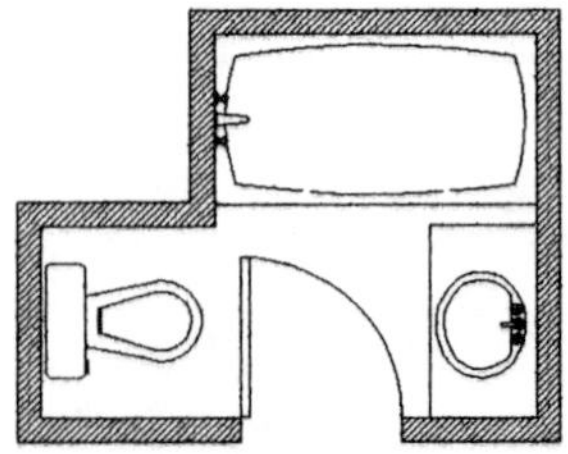

NO. 6 - 38 SF

REMODEL EXISTING BATHROOM	INSTALL NEW BATHROOM
PREPARE PLANS AND OBTAIN PERMITS AS REQUIRED, PERMIT FEE COST **NOT** INCLUDED	PREPARE PLANS AND OBTAIN PERMITS AS REQUIRED, PERMIT FEE COST **NOT** INCLUDED
TEAR OUT EXIST'G BATHROOM FIXTURES & NON-CERAMIC FLOOR, WALL & CEILING COVERINGS	
RE-ROUGH PLUMBING FOR BATHROOM, NEW COPPER SUPPLY AND NEW PLASTIC WASTE LINES TO MAIN STACK WITH CUTOFFS AT EACH FIXTURE INSTALL BUILDER GRADE WHITE FIXTURES, INCLUDING ALL FITTINGS, CAST IRON TUB, BIDET (BATHROOMS 2 AND 5), AND TWO-PIECE TOILET 20" X 18" LAVATORY SET IN TWO-DOOR BUILDER GRADE PREFINISHED VANITY BASE WITH CULTURED MARBLE TOP AND SPLASH	ROUGH IN PLUMBING FOR BATHROOM, COPPER SUPPLY & PLASTIC WASTE LINES TO MAIN STACK WITHIN 5 FEET, WITH CUTOFFS AT EACH FIXTURE INSTALL BUILDER GRADE WHITE FIXTURES, INCLUDING ALL FITTINGS, CAST IRON TUB, BIDET (BATHROOMS 2 AND 5), AND TWO-PIECE TOILET 20" X 18" LAVATORY SET IN TWO-DOOR 36" BUILDER GRADE PREFINISHED VANITY BASE WITH CULTURED MARBLE TOP AND SPLASH
3/4" T & G PLYWOOD SUBFLOOR	
	2-6 X 6-8 BIRCH FLUSH HOLLOW CORE DOOR WITH BATHROOM STYLE LOCKSET
SWITCH AND LIGHT OVER LAVATORY GROUND FAULT DUPLEX OUTLET NEXT TO LAVATORY EXHAUST FAN AND SWITCH VENTED TO OUTSIDE	SWITCH AND LIGHT OVER LAVATORY GROUND FAULT DUPLEX OUTLET NEXT TO LAVATORY EXHAUST FAN VENTED TO OUTSIDE
NEW 1/2" MOISTURE RESISTANT DRYWALL, TAPED AND FINISHED, OVER ALL WALLS AND CEILINGS EXCEPT CERAMIC TILE WALL AREAS NEW CERAMIC TILE ON WALLS OF TUB AREA 6 FT. ABOVE FLOOR, INCL. CERAMIC ACCESSORIES	1/2" MOISTURE RESISTANT DRYWALL, TAPED AND FINISHED, OVER ALL WALLS AND CEILINGS EXCEPT CERAMIC TILE WALL AREAS CERAMIC TILE ON WALLS OF TUB AREA 6 FEET ABOVE FLOOR, INCL. CERAMIC ACCESSORIES
3/8" PLYWOOD UNDERLAYMENT AND NEW CERAMIC TILE FLOOR WITH 4" TILE BASE	3/8" PLYWOOD UNDERLAYMENT AND NEW CERAMIC TILE FLOOR WITH 4" TILE BASE
RECESSED MEDICINE CABINET WITH OVERHEAD FLUORESCENT FIXTURE, 2 MIRROR DOORS, OVERALL SIZE 24" X 16"	RECESSED MEDICINE CABINET WITH OVERHEAD FLUORESCENT FIXTURE, 2 MIRROR DOORS, OVERALL SIZE 24" X 16"
PRIME ROOM AS NECESSARY AND PAINT WITH 2 COATS TOP QUALITY SEMI-GLOSS ENAMEL	PRIME ENTIRE ROOM AND PAINT WITH 2 COATS TOP QUALITY SEMI-GLOSS ENAMEL
REMOVE OLD FIXTURES & TRASH FROM PREMISES & FINAL CLEAN-UP, DUMP FEE **NOT** INCLUDED	REMOVE TRASH FROM PREMISES AND FINAL CLEAN-UP, DUMPING FEE **NOT** INCLUDED

SPECIFICATIONS			UNIT	PRICE	LOCAL AREA MODIFICATION	
					%	PRICE
REMODEL BATHROOM	REMODEL EXISTING BATHROOM	NO. 1	EA	6309.00		
		NO. 2	EA	7884.00		
		NO. 3	EA	6510.00		
		NO. 4	EA	6547.00		
		NO. 5	EA	8481.00		
		NO. 6	EA	6375.00		
INSTALL NEW BATHROOM	NEW BATHROOM	NO. 1	EA	8197.50		
		NO. 2	EA	10,048.50		
		NO. 3	EA	8032.50		
		NO. 4	EA	8059.50		
		NO. 5	EA	10,053.00		
		NO. 6	EA	8346.00		

BATHROOMS WITH STALL SHOWER

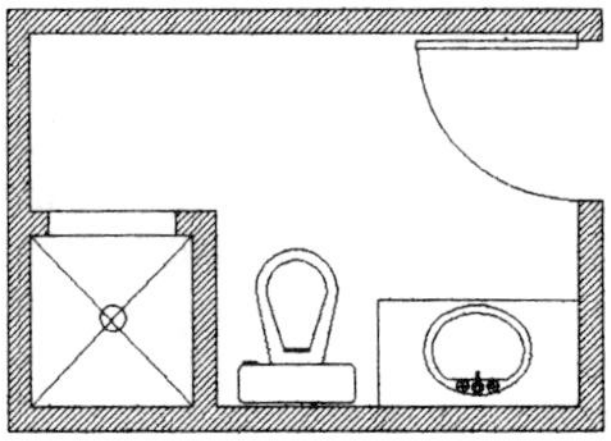

NO. 1 - 50 SF

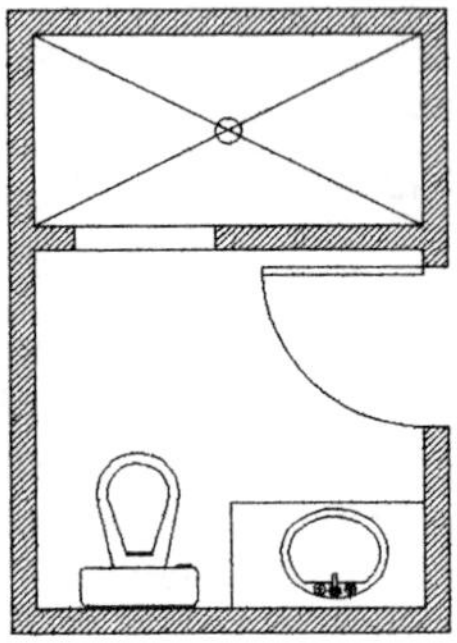

NO. 2 - 54 SF

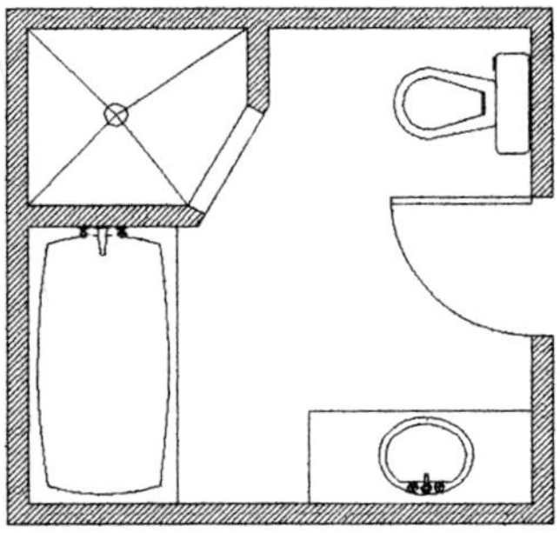

NO. 3 - 77 SF

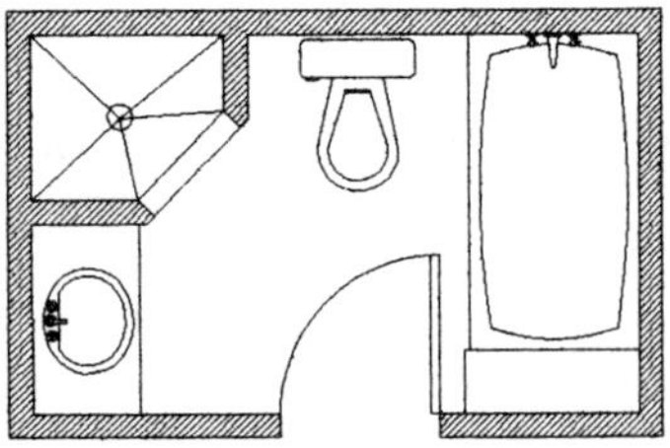

NO. 4 - 57 SF

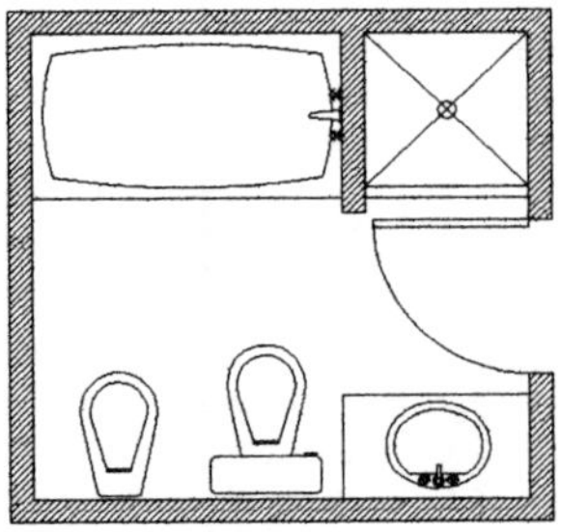

NO. 5 - 60 SF

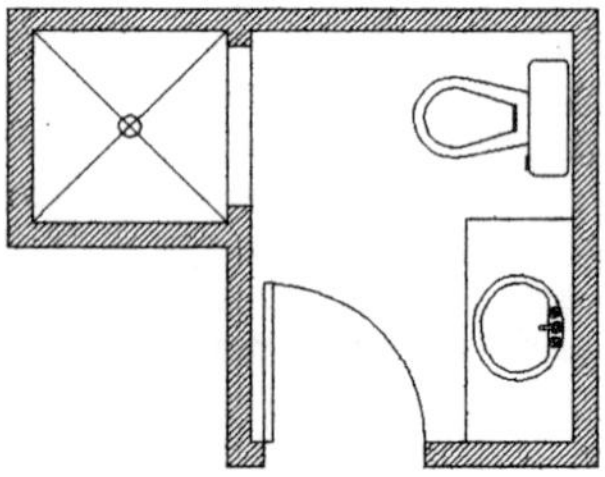

NO. 6 - 40 SF

BATHROOMS WITH STALL SHOWER

REMODEL EXISTING BATHROOM	INSTALL NEW BATHROOM
PREPARE PLANS AND OBTAIN PERMITS AS REQUIRED, PERMIT FEE COST **NOT** INCLUDED	PREPARE PLANS AND OBTAIN PERMITS AS REQUIRED, PERMIT FEE COST **NOT** INCLUDED
TEAR OUT EXIST'G BATHROOM FIXTURES & NON-CERAMIC FLOOR, WALL & CEILING COVERINGS	
RE-ROUGH PLUMBING FOR BATHROOM, NEW COPPER SUPPLY AND NEW PLASTIC WASTE LINES TO MAIN STACK WITH CUTOFFS AT EACH FIXTURE INSTALL BUILDER GRADE WHITE FIXTURES, INCLUDING ALL FITTINGS, CAST IRON TUB, BIDET (BATHROOM 5), TWO-PIECE TOILET AND NEW STALL SHOWER PLUMBING AND FITTINGS, INCLUDING VINYL OR RUBBER PAN ONE OR TWO 20" X 18" LAVATORY BOWLS SET IN BUILDER GRADE PREFINISHED VANITY BASE WITH CULTURED MARBLE TOP AND SPLASH	ROUGH IN PLUMBING FOR BATHROOM, COPPER SUPPLY & PLASTIC WASTE LINES TO MAIN STACK WITHIN 5 FEET, WITH CUTOFFS AT EACH FIXTURE INSTALL BUILDER GRADE WHITE FIXTURES, INCLUDING ALL FITTINGS, CAST IRON TUB, BIDET (BATHROOM 5), TWO-PIECE TOILET AND NEW STALL SHOWER PLUMBING AND FITTINGS, INCLUDING VINYL OR RUBBER PAN ONE OR TWO 20" X 18" LAVATORY BOWLS SET IN BUILDER GRADE PREFINISHED VANITY BASE WITH CULTURED MARBLE TOP AND SPLASH
3/4" T & G PLYWOOD SUBFLOOR	
	2-6 X 6-8 BIRCH FLUSH HOLLOW CORE DOOR WITH BATHROOM STYLE LOCKSET
SWITCH AND LIGHT OVER LAVATORY GROUND FAULT DUPLEX OUTLET NEXT TO LAVATORY EXHAUST FAN AND SWITCH VENTED TO OUTSIDE	SWITCH AND LIGHT OVER LAVATORY GROUND FAULT DUPLEX OUTLET NEXT TO LAVATORY EXHAUST FAN AND SWITCH VENTED TO OUTSIDE
NEW 1/2" MOISTURE RESISTANT DRYWALL, TAPED AND FINISHED, OVER ALL WALLS AND CEILINGS EXCEPT CERAMIC TILE WALL AREAS NEW CERAMIC TILE ON WALLS OF TUB AND SHOWER STALL 6 FEET ABOVE FLOOR, INCLUDING CERAMIC ACCESSORIES	1/2" MOISTURE RESISTANT DRYWALL, TAPED AND FINISHED, OVER ALL WALLS AND CEILINGS EXCEPT CERAMIC TILE WALL AREAS CERAMIC TILE ON WALLS OF TUB AND SHOWER STALL 6 FEET ABOVE FLOOR, INCLUDING CERAMIC ACCESSORIES
3/8" PLYWOOD UNDERLAYMENT AND NEW CERAMIC TILE FLOOR WITH 4" TILE BASE	3/8" PLYWOOD UNDERLAYMENT AND CERAMIC TILE FLOOR WITH 4" TILE BASE
RECESSED MEDICINE CABINET WITH OVERHEAD FLUORESCENT FIXTURE, 2 MIRROR DOORS, OVERALL SIZE 24" X 16"	RECESSED MEDICINE CABINET WITH OVERHEAD FLUORESCENT FIXTURE, 2 MIRROR DOORS, OVERALL SIZE 24" X 16"
PRIME ROOM AS NECESSARY AND PAINT WITH 2 COATS TOP QUALITY SEMI-GLOSS ENAMEL	PRIME ENTIRE ROOM AND PAINT WITH 2 COATS TOP QUALITY SEMI-GLOSS ENAMEL
REMOVE OLD FIXTURES & TRASH FROM PREMISES & FINAL CLEAN-UP, DUMPING FEE **NOT** INCLUDED	REMOVE TRASH FROM PREMISES AND FINAL CLEAN-UP, DUMPING FEE **NOT** INCLUDED

SPECIFICATIONS			UNIT	PRICE	LOCAL AREA MODIFICATION	
					%	PRICE
REMODEL BATHROOM	REMODEL EXISTING BATHROOM	NO. 1	EA	5674.50		
		NO. 2	EA	6850.50		
		NO. 3	EA	8269.50		
		NO. 4	EA	8625.00		
		NO. 5	EA	10,083.00		
		NO. 6	EA	6498.00		
INSTALL NEW BATHROOM	NEW BATHROOM	NO. 1	EA	8305.50		
		NO. 2	EA	9379.50		
		NO. 3	EA	12,390.00		
		NO. 4	EA	11,341.50		
		NO. 5	EA	13,615.50		
		NO. 6	EA	8910.00		

BATHROOMS WITH COMPARTMENTS

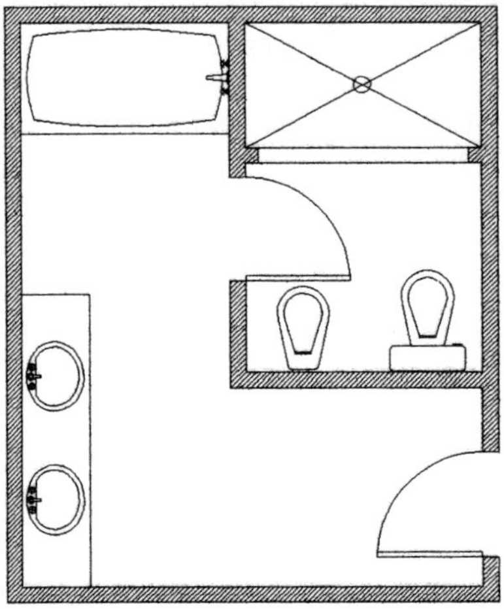

NO. 1 - 148 SF

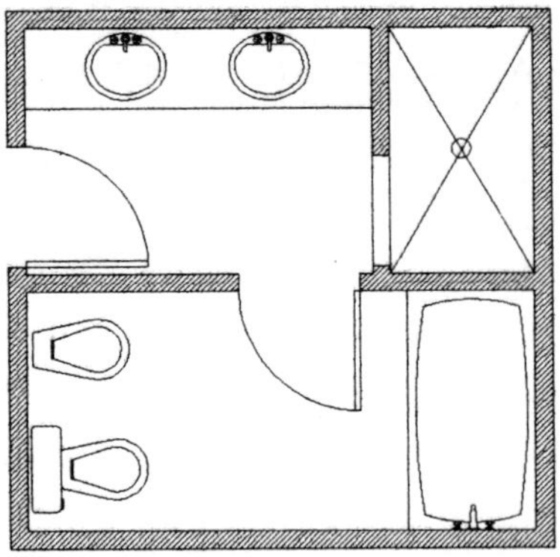

NO. 2 - 110 SF

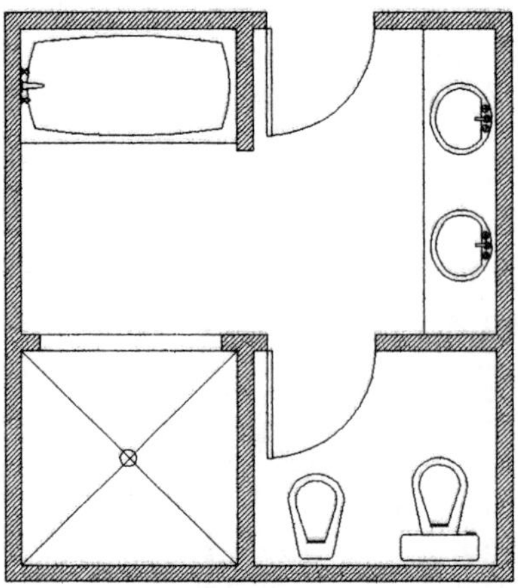

NO. 3 - 136 SF

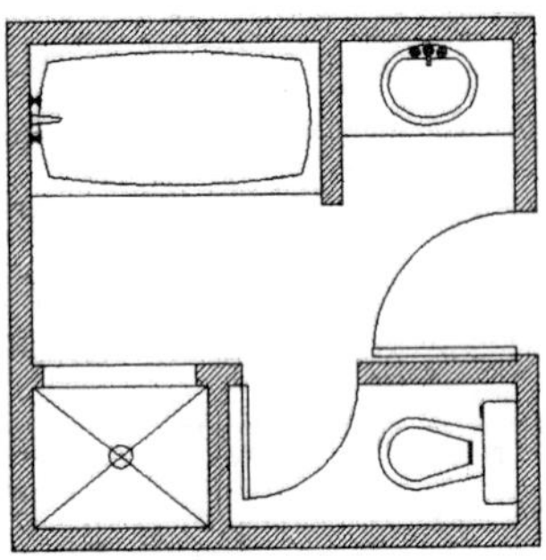

NO. 4 - 70 SF

REMODEL EXISTING BATHROOM	INSTALL NEW BATHROOM
PREPARE PLANS AND OBTAIN PERMITS AS REQUIRED, PERMIT FEE COST **NOT** INCLUDED	PREPARE PLANS AND OBTAIN PERMITS AS REQUIRED, PERMIT FEE COST **NOT** INCLUDED
TEAR OUT EXIST'G BATHROOM FIXTURES & NON-CERAMIC FLOOR, WALL & CEILING COVERINGS	
3/4" T & G PLYWOOD SUBFLOOR	FRAME OUT PARTITIONS WITH 2" X 6", 16" O.C. 3/4" T & G PLYWOOD SUBFLOOR
RE-ROUGH PLUMBING FOR BATHROOM, NEW COPPER SUPPLY AND NEW PLASTIC WASTE LINES TO MAIN STACK WITH CUTOFFS AT EACH FIXTURE INSTALL BUILDER GRADE WHITE FIXTURES, INCLUDING ALL FITTINGS, CAST IRON TUB, BIDET (BATHROOMS 1, 2 AND 3), TWO-PIECE TOILET AND NEW STALL SHOWER PLUMBING AND FITTINGS, INCLUDING VINYL OR RUBBER PAN ONE OR TWO 20" X 18" LAVATORY BOWLS SET IN BUILDER GRADE PREFINISHED VANITY BASE WITH CULTURED MARBLE TOP AND SPLASH	ROUGH IN PLUMBING FOR BATHROOM, COPPER SUPPLY & PLASTIC WASTE LINES TO MAIN STACK WITHIN 5 FEET, CUTOFFS AT EACH FIXTURE INSTALL BUILDER GRADE WHITE FIXTURES, INCLUDING ALL FITTINGS, CAST IRON TUB, BIDET (BATHROOMS 1, 2 AND 3), TWO-PIECE TOILET AND NEW STALL SHOWER PLUMBING AND FITTINGS, INCLUDING VINYL OR RUBBER PAN ONE OR TWO 20" X 18" LAVATORY BOWLS SET IN BUILDER GRADE PREFINISHED VANITY BASE WITH CULTURED MARBLE TOP AND SPLASH
	2-6 X 6-8 BIRCH FLUSH HOLLOW CORE DOOR WITH BATHROOM STYLE LOCKSET
SWITCH AND LIGHT OVER LAVATORY GROUND FAULT DUPLEX OUTLET NEXT TO EACH LAVATORY BOWL EXHAUST FAN AND SWITCH VENTED TO OUTSIDE	SWITCH AND LIGHT OVER LAVATORY GROUND FAULT DUPLEX OUTLET NEXT TO EACH LAVATORY BOWL EXHAUST FAN AND SWITCH VENTED TO OUTSIDE
NEW 1/2" MOISTURE RESISTANT DRYWALL, TAPED AND FINISHED, OVER ALL WALLS AND CEILINGS EXCEPT CERAMIC TILE WALL AREAS NEW CERAMIC TILE ON WALLS OF TUB AND SHOWER STALL 6 FEET ABOVE FLOOR, INCLUDING CERAMIC ACCESSORIES	1/2" MOISTURE RESISTANT DRYWALL, TAPED AND FINISHED, OVER ALL WALLS AND CEILINGS EXCEPT CERAMIC TILE WALL AREAS CERAMIC TILE ON WALLS OF TUB AND SHOWER STALL 6 FEET ABOVE FLOOR, INCLUDING CERAMIC ACCESSORIES
3/8" PLYWOOD UNDERLAYMENT AND NEW CERAMIC TILE FLOOR WITH 4" TILE BASE	3/8" PLYWOOD UNDERLAYMENT AND CERAMIC TILE FLOOR WITH 4" TILE BASE
RECESSED MEDICINE CAB'T W/OVERHEAD FLUORESCENT FIXTURE, 2 MIRROR DOORS, OVERALL SIZE 24" X 16", OVER EACH LAVATORY BOWL	RECESSED MEDICINE CAB'T W/OVERHEAD FLUORESCENT FIXTURE, 2 MIRROR DOORS, OVERALL SIZE 24" X 16", OVER EACH LAVATORY BOWL
PRIME ROOM AS NECESSARY AND PAINT WITH 2 COATS TOP QUALITY SEMI-GLOSS ENAMEL	PRIME ENTIRE ROOM AND PAINT WITH 2 COATS TOP QUALITY SEMI-GLOSS ENAMEL
REMOVE OLD FIXTURES & TRASH FROM PREMISES & FINAL CLEAN-UP, DUMPING FEE **NOT** INCLUDED	REMOVE TRASH FROM PREMISES AND FINAL CLEAN-UP, DUMPING FEE **NOT** INCLUDED

SPECIFICATIONS			UNIT	PRICE	LOCAL AREA MODIFICATION	
					%	PRICE
REMODEL BATHROOM	REMODEL EXISTING BATHROOM	NO. 1	EA	15,280.50		
		NO. 2	EA	13,704.00		
		NO. 3	EA	15,757.50		
		NO. 4	EA	9,904.50		
INSTALL NEW BATHROOM	NEW BATHROOM	NO. 1	EA	20,172.00		
		NO. 2	EA	18,342.00		
		NO. 3	EA	20,638.50		
		NO. 4	EA	13,423.50		

POWDER ROOMS

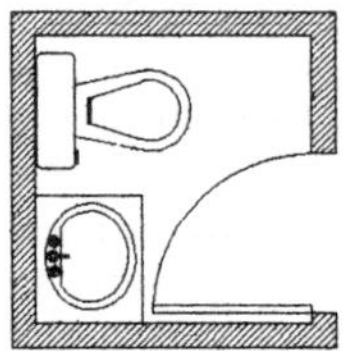

NO. 1 - 20 SF

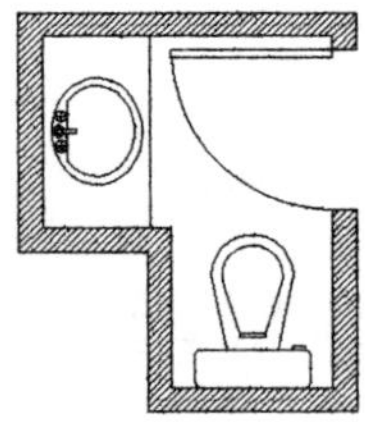

NO. 2 - 20 SF

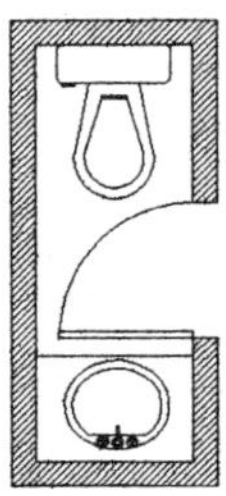

NO. 3 - 20 SF

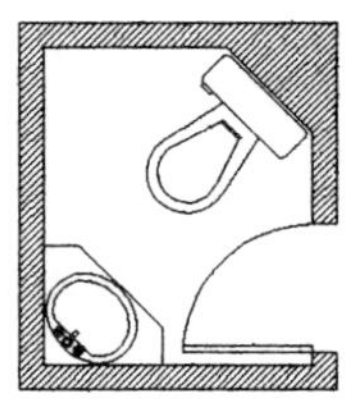

NO. 4 - 16 SF

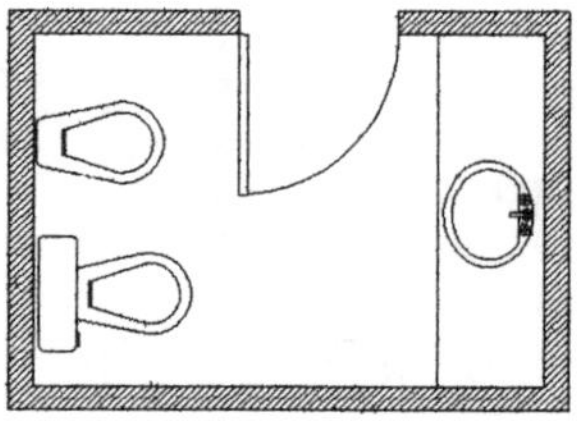

NO. 5 - 44 SF

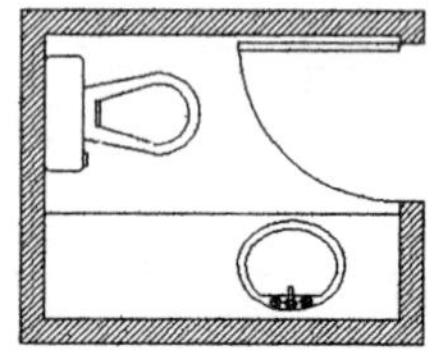

NO. 6 - 25 SF

POWDER ROOMS

REMODEL EXISTING POWDER ROOM	INSTALL NEW POWDER ROOM
REPARE PLANS AND OBTAIN PERMITS AS EQUIRED, PERMIT FEE COST **NOT** INCLUDED	PREPARE PLANS AND OBTAIN PERMITS AS REQUIRED, PERMIT FEE COST **NOT** INCLUDED
EAR OUT EXIST'G POWDER ROOM FIXTURES & ON-CERAMIC FLOOR, WALL & CEILING COVERINGS	
E-ROUGH PLUMBING FOR POWDER ROOM, NEW OPPER SUPPLY AND NEW PLASTIC WASTE LINES O MAIN STACK WITH CUTOFFS AT EACH FIXTURE	ROUGH IN PLUMB'G FOR POWDER ROOM, COPPER SUPPLY & PLASTIC WASTE LINES TO MAIN STACK WITHIN 5 FT, W/CUTOFFS AT EACH FIXTURE
ISTALL BUILDER GRADE WHITE FIXTURES, ICLUDING ALL FITTINGS, BIDET (POWDER ROOM), AND TWO-PIECE TOILET	INSTALL BUILDER GRADE WHITE FIXTURES, INCLUDING ALL FITTINGS, BIDET (POWDER ROOM 5), AND TWO-PIECE TOILET
0" X 18" LAVATORY BOWL SET IN PREFINISHED ANITY BASE W/IMITATION MARBLE TOP & SPLASH	20" X 18" LAVATORY BOWL SET IN PREFINISHED VANITY BASE W/IMITATION MARBLE TOP & SPLASH
/4" T & G PLYWOOD SUBFLOOR	
	2-6 X 6-8 BIRCH FLUSH HOLLOW CORE DOOR WITH BATHROOM STYLE LOCKSET
WITCH AND LIGHT OVER LAVATORY ROUND FAULT DUPLEX OUTLET NEXT TO AVATORY XHAUST FAN AND SWITCH VENTED TO OUTSIDE	SWITCH AND LIGHT OVER LAVATORY GROUND FAULT DUPLEX OUTLET NEXT TO LAVATORY EXHAUST FAN AND SWITCH VENTED TO OUTSIDE
EW 1/2" DRYWALL (GYPSUM BOARD), TAPED AND INISHED, OVER ALL WALLS AND CEILING	1/2" DRYWALL (GYPSUM BOARD), TAPED AND FINISHED, OVER ALL WALLS AND CEILING
/8" PLYWOOD UNDERLAYMENT AND NEW ERAMIC TILE FLOOR WITH 4" TILE BASE	3/8" PLYWOOD UNDERLAYMENT AND CERAMIC TILE FLOOR WITH 4" TILE BASE
ECESSED MEDICINE CABINET WITH OVERHEAD LUORESCENT FIXTURE, 2 MIRROR DOORS, VERALL SIZE 24" X 16"	RECESSED MEDICINE CABINET WITH OVERHEAD FLUORESCENT FIXTURE, 2 MIRROR DOORS, OVERALL SIZE 24" X 16"
RIME ROOM AS NECESSARY AND PAINT WITH 2 OATS TOP QUALITY SEMI-GLOSS ENAMEL	PRIME ENTIRE ROOM AND PAINT WITH 2 COATS TOP QUALITY SEMI-GLOSS ENAMEL
EMOVE OLD FIXTURES & TRASH FROM PREMISES FINAL CLEAN-UP, DUMPING FEE **NOT** INCLUDED	REMOVE TRASH FROM PREMISES AND FINAL CLEAN-UP, DUMPING FEE **NOT** INCLUDED

SPECIFICATIONS			UNIT	PRICE	LOCAL AREA MODIFICATION %	LOCAL AREA MODIFICATION PRICE
EMODEL OWDER OOM	REMODEL EXISTING POWDER ROOM	NO. 1	EA	2917.50		
		NO. 2	EA	2770.50		
		NO. 3	EA	2916.00		
		NO. 4	EA	2952.00		
		NO. 5	EA	6108.00		
		NO. 6	EA	4074.00		
NSTALL NEW OWDER OOM	NEW POWDER ROOM	NO. 1	EA	4501.50		
		NO. 2	EA	4467.00		
		NO. 3	EA	4635.00		
		NO. 4	EA	4635.00		
		NO. 5	EA	8661.00		
		NO. 6	EA	5721.00		

BATHROOMS & POWDER ROOMS — EXTRAS & ALLOWANCES

SPECIFICATIONS		UNIT	PRICE	LOCAL AREA MODIFICATION	
				%	PRICE
ADDITIONAL AREA	ADDITIONAL FLOOR, WALL AND CEILING AREA TO BATHROOM OR POWDER ROOM (ADDITIONAL FIXTURES MAY REQUIRE ADDITIONAL FLOOR SPACE) SF = FLOOR AREA OF ADDITIONAL SPACE	SF	28.14		
FIXTURES	SUBSTITUTE ONE-PIECE LUXOR OR EQUAL W.C. WITH VENTAWAY FEATURE, WHITE **ADD**	EA	450.00		
	SUBSTITUTE STEEL BATHTUB, FOR SPECIFIED CAST IRON TUB **DEDUCT**	EA	309.00		
	FIBERGLASS WHIRLPOOL TUB, 60" X 30" X 18" **ADD**	EA	1944.00		
	SUBSTITUTE PEDESTAL SINK, VITREOUS CHINA, 24" X 19" FOR ANY 36" VANITY BASE WITH LAVATORY SET IN **DEDUCT**	EA	324.00		
	ADD OR OMIT BUILDER GRADE BIDET **ADD** OR **DEDUCT**	EA	1455.00		
	SUBSTITUTE 36" VANITY BASE AND ONE LAVATORY BOWL FOR SPECIFIED 84" VANITY BASE AND TWO LAVATORY BOWLS **DEDUCT**	EA	1365.00		
	GOING AWAY FROM STACK ON ANY PLUMBING FIXTURE INSTALLATION **ADD** LF = EACH FIXTURE, TOTAL DISTANCE FROM EXISTING STACK	LF	54.00		

BATHROOMS & POWDER ROOMS — EXTRAS & ALLOWANCES

SPECIFICATIONS			UNIT	PRICE	LOCAL AREA MODIFICATION	
					%	PRICE
ELECTRICAL	ADDITIONAL DUPLEX OUTLETS					
	IN OPEN FRAMING	**ADD**	EA	57.00		
	FISHED IN FRAME WALL	**ADD**	EA	96.00		
	FISHED IN MASONRY WALL	**ADD**	EA	132.00		
	ADDITIONAL EXHAUST FAN IN COMPARTMENT BATH	**ADD**	EA	204.00		
	HEAT LIGHT, INFRA-RED	**ADD**	EA	324.00		
	ADDITIONAL GFI OUTLET	**ADD**	EA	117.00		
	TELEPHONE JACK	**ADD**	EA	156.00		
FLOOR COVERING	SUBSTITUTE SHEET VINYL FLOOR COVERING, INCLUDING VINYL BASE AND CORNERS, FOR SPECIFIED CERAMIC TILE FLOOR SF = FLOOR	**DEDUCT**	SF	9.84		

BREEZEWAY BETWEEN EXISTING HOUSE AND GARAGE

SPECIFICATIONS	
PLANS AND PERMIT	PREPARE PLANS AND OBTAIN PERMITS AS REQUIRED, PERMIT FEE COST **NOT** INCLUDED
CONCRETE AND MASONRY	8" X 16" CONTINUOUS CONCRETE FOOTINGS, 36" BELOW GRADE 8" X 8" X 16" BLOCK BUILDUP TO 12" ABOVE GRADE 4" CONCRETE FLOOR SLAB OVER GRAVEL WITH NO. 4 (1/2") REBAR INTO EXISTING HOUSE FOUNDATION WALL, 24" O.C.
FRAMING	4" X 4" PRESSURE TREATED PINE OR FIR POSTS, 6'-0" O.C. 4" X 10" PRESSURE TREATED HEADERS GABLE TYPE ROOF WITH 12" OVERHANG, 4/12 SLOPE, INCLUDING CEILING JOISTS 1/2" CDX PLYWOOD ROOF SHEATHING
ROOF COVERING, FLASHING, GUTTERS	#215 FIBERGLASS STRIP SHINGLES ALUMINUM STEP FLASHING TO EXISTING HOUSE AND GARAGE OR CARPORT WALLS ALUMINUM GUTTERS AND DOWNSPOUTS
EXTERIOR TRIM	6" TIGHT KNOT PINE OR FIR FASCIA 12" TIGHT KNOT PINE OR FIR SOFFIT 5/8" X 4" FIR BEADED CEILING 3/4" CEILING COVE MOULDING AT HEADER INTERSECTIONS
CLEAN-UP	REMOVE TRASH FROM PREMISES, DUMPING FEE NOT INCLUDED
	NO PAINTING

SPECIFICATIONS			UNIT	PRICE	LOCAL AREA MODIFICATION %	LOCAL AREA MODIFICATION PRICE
BREEZEWAY	AS SPECIFIED		EA PLUS SF	987.00 25.53		
DIFFERENT FOOTING DEPTH	***Depth of Bottom of Footing Below Grade***					
	12"	**DEDUCT**	LF	20.94		
	24"	**DEDUCT**	LF	10.46		
	48"	**ADD**	LF	10.43		
	LF = FOOTINGS					

SPECIFICATIONS	
PLANS AND PERMIT	PREPARE PLANS AND OBTAIN PERMITS AS REQUIRED, PERMIT FEE COST **NOT** INCLUDED
CONCRETE	MONOLITHIC COMBINATION FOOTING AND 4" CONCRETE FLOOR SLAB, REINFORCED, 36" BELOW GRADE, WITH NO. 4 (1/2") REBAR INTO EXISTING HOUSE FOUNDATION WALL, 24" O.C.
FRAMING	PRESSURE TREATED 4" X 4" PINE OR FIR POSTS, 8'-0" O.C., 4" X 10" PRESSURE TREATED HEADERS GABLE TYPE ROOF WITH 12" OVERHANG, 4/12 SLOPE, **NO** OVERLAY, RAFTERS AND CEILING JOISTS 16" O.C. 1/2" PLYWOOD ROOF SHEATHING
ROOFING, GUTTERS, FLASHING	#215 FIBERGLASS SHINGLES OVER #15 FELT PAPER TWO ALUMINUM GUTTERS AND DOWNSPOUTS ALUMINUM FLASHING AT EXISTING HOUSE
EXTERIOR TRIM	TRIM HEADERS 3 SIDES WITH TIGHT KNOT PINE OR FIR TIGHT KNOT PINE OR FIR FASCIA AND SOFFIT RAKE AND RAKE MOULDING ON GABLE END 5/8" X 4" FIR BEADED CEILING 3/4" CEILING COVE MOULDING AT HEADER INTERSECTIONS
SIDING	8" HORIZONTAL VINYL SIDING, UNINSULATED, ON GABLE END(S)
CLEAN-UP	REMOVE TRASH FROM PREMISES, DUMPING FEE **NOT** INCLUDED
	NO PAINTING

SPECIFICATIONS				UNIT	PRICE	LOCAL AREA MODIFICATION %	LOCAL AREA MODIFICATION PRICE
ATTACHED CARPORT	AS SPECIFIED			EA PLUS SF	2670.00 21.75		
EXTRAS AND ALLOWANCES	***Depth of Bottom of Footing Below Grade***						
		12"	**DEDUCT**	LF	3.84		
		24"	**DEDUCT**	LF	1.91		
		48"	**ADD**	LF	1.95		
	LF = FOOTINGS						

DECK

SPECIFICATIONS	
PLANS AND PERMIT	PREPARE PLANS AND OBTAIN PERMITS AS REQUIRED, PERMIT FEE COST **NOT** INCLUDED
CONCRETE	DIG OUT 12" X 12" HOLES TO 36" BELOW GRADE, 8'-0" O.C. BUILD FORMS WITH DIMENSION LUMBER FROM GRADE TO 6" ABOVE GRADE POUR CONCRETE INTO EXCAVATION TO TOP OF FORMS AND SET METAL POST ANCHORS IN CONCRETE
FRAMING & DECK	4" X 4" WOOD POSTS, 8'-0" O.C. 2" X 8" OR 2" X 10" LEDGER BOARD BOLTED TO HOUSE WITH NO. 4 (1/2") LAG BOLTS, 16" O.C. (2) 2" X 10" WOOD BEAM AND 2" X 8" OR 2" X 10" JOISTS 16" O.C., CANTILEVERED 24", 2" X 8" OR 2" X 10" BAND 2" X 6" DECK LAID FLAT WITH RUST RESISTANT NAILS SPACED 1/4" APART
CLEAN-UP	REMOVE TRASH FROM PREMISES, DUMPING FEE COST **NOT** INCLUDED
	NO PAINTING OR STAINING

SPECIFICATIONS		UNIT	PRICE	LOCAL AREA MODIFICATION	
				%	PRICE
WOOD DECK	BUILT WITH PRESSURE TREATED FIR OR PINE FRAMING AND 2" X 6" FIR OR PINE DECK SURFACE	SF	10.65		
	BUILT WITH PRESSURE TREATED FIR OR PINE FRAMING, 1" X 10" CEDAR OR REDWOOD BAND AND 2" X 6" CEDAR OR REDWOOD DECK SURFACE	SF	15.00		
	BUILT WITH PRESSURE TREATED FIR OR PINE FRAMING, 1" X 10" CEDAR OR REDWOOD BAND AND 5/4" X 6" CEDAR OR REDWOOD DECK SURFACE	SF	12.71		
ALTERNATE DECK FOOTINGS SYSTEM	DIG OUT 20" X 20" HOLES TO 36" BELOW GRADE, 8'-0" O.C. AND POUR 8" THICK CONCRETE FOOTINGS. BUILD BLOCK PIER TO 8" ABOVE GRADE AND FILL WITH CONCRETE WITH METAL POST ANCHORS SET IN CONCRETE **ADD**	EA	28.50		

SPECIFICATIONS	
DECK RAILING NO. 1	ATTACH 4" X 4" END AND INTERMEDIATE POSTS TO BAND EXTENDING TO 42" ABOVE SURFACE OF DECK, 4'-0" O.C. 2" X 6" CAP LAID FLAT ON POSTS, 2" X 6" TOP RAIL UNDER CAP AND 2" X 6" MIDDLE AND BOTTOM RAILS ATTACHED TO THE OUTSIDE OF POSTS ON THE DECK SIDE
DECK RAILING NO. 2	ATTACH 4" X 4" END AND INTERMEDIATE POSTS TO BAND EXTENDING TO 42" ABOVE SURFACE OF DECK, 4'-0" O.C. 2" X 6" CAP LAID FLAT ON POSTS, 2" X 6" TOP RAIL UNDER CAP AND 2" X 2" PICKETS 7-1/2" O.C. NAILED TO OUTSIDE TOP RAIL AND BAND
STEPS	2" X 12" STRINGERS TWO 2" X 6" PER TREAD, OPEN RISERS, CONCRETE BOTTOM TREAD ON GROUND STEP RAILING ON 2 SIDES WITH CAP LAID FLAT ON POSTS, 2" X 6" TOP RAIL UNDER CAP AND 2" X 6" MIDDLE AND BOTTOM RAILS ATTACHED TO THE OUTSIDE OF POSTS ON STEP SIDE

SPECIFICATIONS		UNIT	PRICE	LOCAL AREA MODIFICATION	
				%	PRICE
DECK RAILING NO. 1	BUILT WITH PRESSURE TREATED FIR OR PINE	LF	12.30		
	BUILT WITH CEDAR OR REDWOOD	LF	16.76		
DECK RAILING NO. 2	BUILT WITH PRESSURE TREATED FIR OR PINE	LF	12.60		
	BUILT WITH CEDAR OR REDWOOD	LF	17.18		
STEPS	36" WIDE STEPS BUILT WITH PRESSURE TREATED FIR OR PINE	PER STEP	40.50		
	36" WIDE STEPS BUILT WITH CEDAR OR REDWOOD	PER STEP	64.50		
	60" WIDE STEPS BUILT WITH PRESSURE TREATED FIR OR PINE	PER STEP	58.50		
	60" WIDE STEPS BUILT WITH CEDAR OR REDWOOD	PER STEP	93.00		

SHELL DORMER

NOTE: TO OBTAIN THE SQUARE FOOT MEASUREMENT OF DORMER, FIRST *MEASURE THE DISTANCE FROM A POINT DIRECTLY BELOW THE RIDGEBOARD TO THE OUTSIDE WALL OF THE FUTURE DORMER.* THE OUTSIDE WALL OF THE FUTURE DORMER WILL USUALLY BE EVEN WITH THE OUTSIDE WALL OF THE EXISTING HOUSE, OR IT MAY BE STEPPED IN 1 OR 2 FEET FROM THE EXISTING OUTSIDE WALL. *MULTIPLY THAT MEASUREMENT BY THE WIDTH OF THE DORMER* (THE DISTANCE FROM OUTSIDE TO OUTSIDE OF THE SIDE WALLS OF THE FUTURE DORMER). THE RESULT IS THE SQUARE FOOT AREA OF THE FUTURE DORMER.

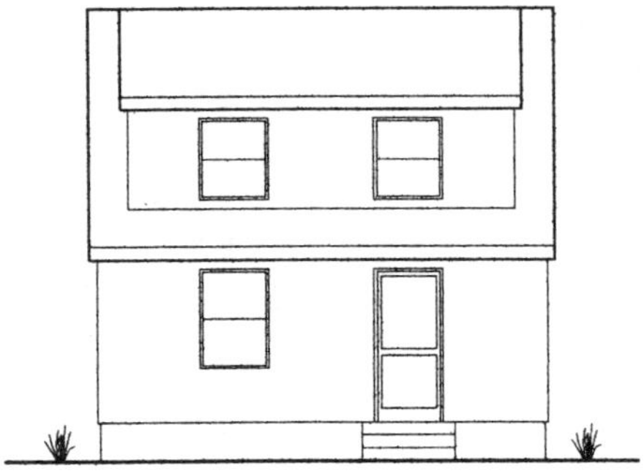

NO.1 STEPPED IN FRONT AND SIDES

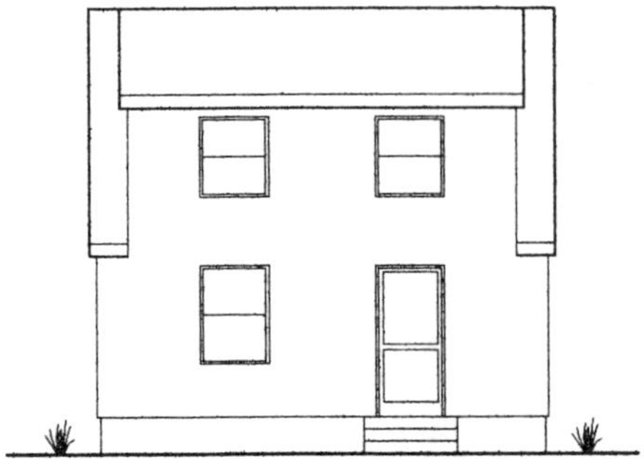

NO.2 STEPPED IN SIDES, FLUSH FRONT

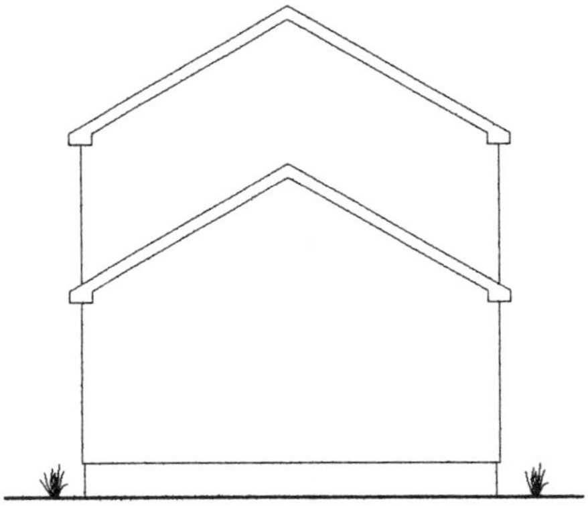

NO.3 ADD A LEVEL -- STEPPED IN ALL AROUND

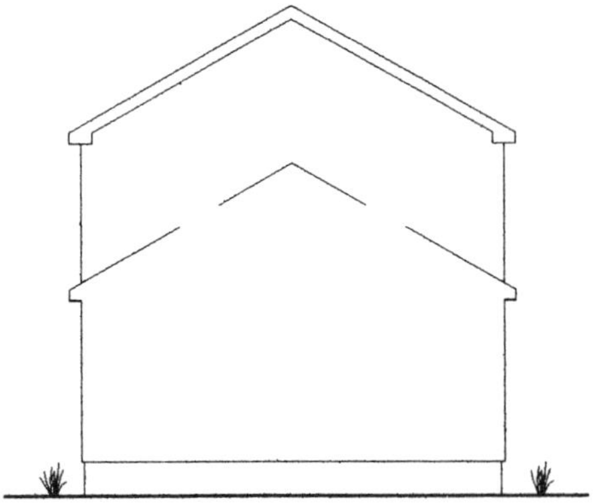

NO.4 ADD A LEVEL -- FLUSH ALL AROUND

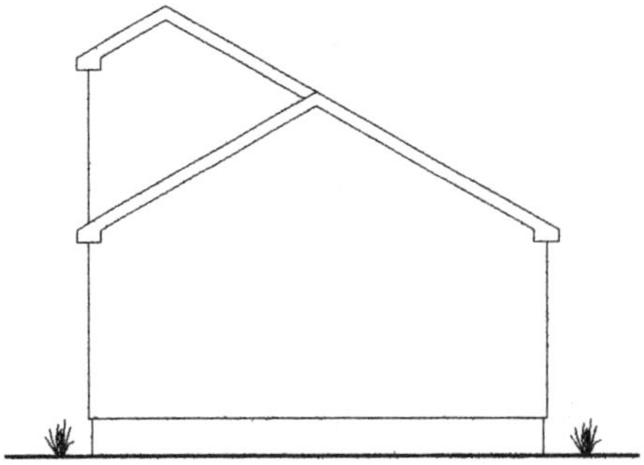

NO.5 UP AND OVER -- STEPPED IN FRONT AND SIDES

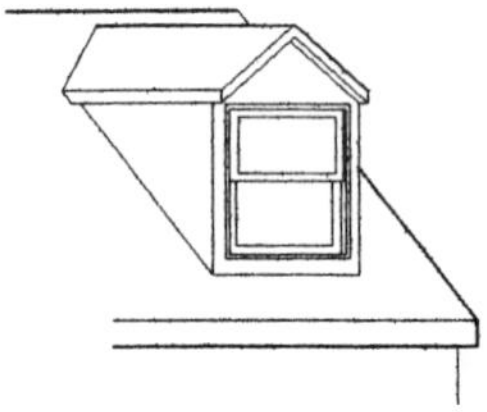

NO.6 SMALL GABLE (DOGHOUSE) DORMER

SPECIFICATIONS	
PLANS AND PERMITS	PREPARE PLANS AND OBTAIN PERMITS AS REQUIRED, PERMIT FEE COST **NOT** INCLUDED
TEAR-OUT	REMOVE EXISTING ROOF COVERING, RAFTERS AND SHEATHING WHERE NECESSARY
WALL FRAMING	2" X 4" STUDS, 16" O.C. 1/2" PLYWOOD SHEATHING
ROOF FRAMING	SHED TYPE ROOF WITH 12" OVERHANG RAFTERS AND CEILING JOISTS, 16" O.C., FRAMED INTO EXISTING RIDGE BOARD 1/2" PLYWOOD SHEATHING
ROOFING, GUTTERS, FLASHING	#215 FIBERGLASS STRIP SHINGLES OVER #15 FELT PAPER ALUMINUM GUTTERS AND DOWNSPOUTS ALUMINUM FLASHING WHERE DORMER JOINS EXISTING ROOF
EXTERIOR TRIM	#2 PINE FASCIA AND SOFFIT #2 PINE RAKE WITH RAKE MOULDING SOFFIT VENTS OR GABLE VENTS AS REQUIRED
SIDING	VINYL SIDING, INSULATED, 8" HORIZONTAL
WINDOWS	*DORMERS 1, 2 AND 5:* 3 WOOD DOUBLE HUNG 2-8 X 3-10 WINDOWS, DOUBLE GLAZED *DORMERS 3 AND 4:* 6 WOOD DOUBLE HUNG 2-8 X 3-10 WINDOWS, DOUBLE GLAZED *DORMER 6:* 1 WOOD DOUBLE HUNG 2-8 X 3-10 WINDOW, DOUBLE GLAZED
CLEAN-UP	REMOVE TRASH FROM PREMISES, DUMPING FEE **NOT** INCLUDED
	NO PAINTING

SPECIFICATIONS		UNIT	PRICE	LOCAL AREA MODIFICATION	
				%	PRICE
SHELL DORMER	SF = SHELL DORMER DIMENSIONS (SEE PAGES 34 AND 38)				
	NO. 1 STEPPED IN FRONT AND SIDES	EA PLUS SF	3090.00 16.44		
	NO. 2 STEPPED IN SIDES, FLUSH FRONT	EA PLUS SF	3090.00 17.97		
	NO. 3 ADD A LEVEL — STEPPED IN ALL AROUND	EA PLUS SF	6825.00 15.80		
	NO. 4 ADD A LEVEL — FLUSH ALL AROUND	EA PLUS SF	6825.00 18.87		
	NO. 5 UP AND OVER — STEPPED IN FRONT AND SIDES	EA PLUS SF	1719.00 16.07		
	NO. 6 GABLE (DOGHOUSE) DORMER (**NO** GUTTER AND DOWNSPOUT)	EA	1530.00		

SHELL DORMER — EXTRAS AND ALLOWANCES

SPECIFICATIONS		UNIT	PRICE	LOCAL AREA MODIFICATION %	LOCAL AREA MODIFICATION PRICE
THIRD LEVEL DORMER	DORMER INSTALLED ON 3rd STORY ABOVE GRADE **ADD** SF = DORMER DIMENSION	SF	3.45		
DORMER ON SLATE ROOF	CUT INTO EXISTING SLATE ROOF AND INSTALL DORMER **ADD** SF = DORMER DIMENSION	SF	1.20		
ROOF FRAMING	SUBSTITUTE GABLE STYLE ROOF FOR SPECIFIED SHED TYPE ROOF **ADD** SF = DORMER ROOF	SF	1.80		
ROOF COVERING	SUBSTITUTE ROOF COVERINGS FOR SPECIFIED FIBERGLASS SHINGLES				
	CONCRETE ROOFING TILES **ADD**	SF	2.31		
	18" CEDAR SHINGLES **ADD**	SF	4.50		
	SF = DORMER ROOF				
TEAR-OUT	REMOVE EXISTING NON-BEARING PARTITION	LF	9.96		
	REMOVE EXISTING INSULATION AND CEILING COVERING FROM RAFTERS SF = SF RAFTERS	SF	.57		
WINDOWS	ADD OR OMIT ONE WOOD DOUBLE HUNG WINDOW, 2-8 X 3-10, DOUBLE GLAZED **ADD** OR **DEDUCT**	EA	348.00		
	SUBSTITUTE DOUBLE GLAZED 4-0 X 4-0 ALUMINUM SLIDING WINDOW **ADD** OR **DEDUCT**	EA	105.00		
BREAK THROUGH AND INSTALL WINDOW	BREAK THROUGH EXISTING GABLE END, INSTALL HEADER AND/OR LINTEL AS REQUIRED AND INSTALL WINDOW, INCLUDING SASH, FRAME AND ALL TRIM, WOOD DOUBLE HUNG, DOUBLE GLAZED, 2-8 X 3-10				
	FRAME WALL	EA	795.00		
	MASONRY WALL	EA	840.00		

SHELL DORMER — EXTRAS AND ALLOWANCES

SPECIFICATIONS		UNIT	PRICE	LOCAL AREA MODIFICATION	
				%	PRICE
FLOOR FRAMING	SISTER (DOUBLE UP) EXISTING 2nd FLOOR JOISTS				
	2" X 6" **ADD**	SF	3.57		
	2" X 8" **ADD**	SF	4.10		
	SF = FLOOR JOISTS				
	REMOVE EXISTING FLOOR JOISTS, SUBFLOOR AND CEILING BELOW AND INSTALL HEADER FOR STAIRS	EA	429.00		
KNEEWALL	INTERIOR KNEEWALL, 4 TO 6 FEET HIGH, 2" X 4" 16" O.C.	LF	12.44		
FURRING	FURRING APPLIED OVER FRAME WALLS OR STRAIGHT MASONRY WALL	SF	.92		
	FURRING APPLIED OVER CROOKED MASONRY WALLS	SF	1.16		
FRAME PARTITION	FRAME OUT 2" X 4" WALL 8'-0" HIGH	LF	16.65		
EXTEND CHIMNEY	EXTEND EXISTING CHIMNEY ABOVE ROOF				
	1 FLUE	LF UP	108.00		
	2 FLUES	LF UP	135.00		
SIDING	SUBSTITUTE OTHER SIDINGS FOR SPECIFIED VINYL SIDING				
	BEVELED CEDAR **ADD**	SF	.86		
	18" CEDAR SHINGLES **ADD**	SF	.54		
	SF = DORMER WALLS				
RIDGE BOARD	DOUBLE UP RIDGE BOARD				
	2" X 10" **ADD**	LF	3.08		
	2" X 12" **ADD**	LF	3.90		

DORMER INTERIOR TRIM AND FINISHING

THE DIMENSIONS FOR INTERIOR TRIM AND FINISHING WILL USUALLY BE MORE THAN THE SHELL DORMER DIMENSION AS THE NEW LIVING AREA WILL PROBABLY EXTEND PAST THE EXISTING RIDGE-BOARD AS FAR AS ADEQUATE HEAD-ROOM CAN BE OBTAINED.

BE SURE TO INCLUDE ANY LIVING AREA AT EACH SIDE OF THE DORMER INTERIOR IF THE INSTALLATION IS STEPPED IN SIDES.

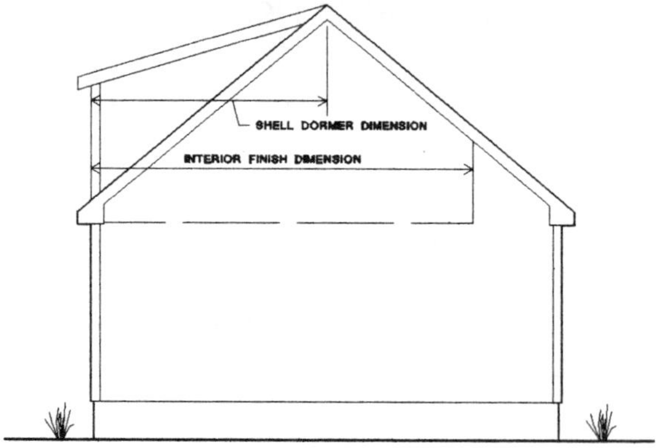

SPECIFICATIONS

WINDOWS	TRIM WINDOWS WITH 2-1/4" CASING, STOOLS, STOPS AND SASH FASTENERS
HEATING	EXTEND HEAT AND RETURN DUCTS FROM EXISTING SERVICE
ELECTRIC	DUPLEX WALL OUTLETS TO CODE ON EXISTING SERVICE
INSULATION	3-1/2" FIBERGLASS BLANKET INSULATION ON ALL SIDEWALLS 6" FIBERGLASS BLANKET INSULATION IN CEILING
INTERIOR WALL COVERING	1/2" GYPSUM BOARD (SHEETROCK) ON THREE NEW WALLS AND ON FURRING STRIPS OVER EXISTING HOUSE WALL, TAPED AND FINISHED (**NO** PAINTING)
CEILING COVERING	1/2" GYPSUM BOARD (SHEETROCK) TAPED AND FINISHED (**NO** PAINTING)
INTERIOR TRIM	3-1/2" BASE AND SHOE MOULDING ON ALL FOUR WALLS
FLOOR COVERING	2-1/4" SELECT OAK FLOORING, SANDED AND FINISHED
CLEAN-UP	REMOVE RUBBISH FROM PREMISES, DUMPING FEE **NOT** INCLUDED
	NO PAINTING

SPECIFICATIONS		UNIT	PRICE	LOCAL AREA MODIFICATION	
				%	PRICE
INTERIOR TRIM AND FINISHING	AS SPECIFIED SF = TOTAL SQUARE FOOT LIVING AREA TO BE FINISHED	EA PLUS SF	1530.00 17.40		

DORMER INTERIOR TRIM & FINISH — EXTRAS & ALLOWANCES

SPECIFICATIONS		UNIT	PRICE	LOCAL AREA MODIFICATION %	LOCAL AREA MODIFICATION PRICE
PARTITION	TRIM AND FINISH FRAMED OUT PARTITION: • DUPLEX OUTLETS TO CODE BOTH SIDES OF WALL • 1/2" DRYWALL BOTH SIDES OF WALL, FINISHED BUT **NOT** PAINTED • 3-1/2" BASE AND SHOE BOTH SIDES OF WALL LF = LENGTH OF PARTITION	LF	42.00		
DOOR	HANG DOOR IN FRAMED OUT PARTITION, 2-6 X 6-8, BIRCH HOLLOW CORE, INCLUDING JAMBS, STOPS, CASINGS AND HARDWARE **ADD**	EA	252.00		
	BI-FOLD DOORS, BIRCH FLUSH, INCLUDING ALL TRIM AND HARDWARE				
	2 DOORS 3-0 X 6-8 **ADD**	SET	216.00		
	6-0 X 6-8 **ADD**	SET	264.00		
	4 DOORS 4-0 X 6-8 **ADD**	SET	255.00		
	6-0 X 6-8 **ADD**	SET	285.00		
WINDOW	TRIM ADDITIONAL WINDOW OR OMIT TRIMMING WINDOW **ADD** OR **DEDUCT**	EA	75.00		
CLOSET TRIM	TRIM FRAMED CLOSET WITH BASE AND SHOE, HOOK-STRIP, SHELF, CLOTHES POLE AND CLOTHES POLE SOCKET	LF	23.67		
CEILING MOULDING	CEILING MOULDING AT WALL INTERSECTIONS				
	3/4" **ADD**	LF	1.76		
	1-5/8" **ADD**	LF	2.28		
	3-5/8" **ADD**	LF	3.78		
STAIRS	CLOSED STAIRWAY TO 2nd FLOOR DORMER ROOM(S), OAK TREADS AND RISERS WITH BIRCH HANDRAIL, SHOP BUILT AND INSTALLED ON JOB BY STAIRBUILD-ER EA = EACH STAIRWAY	EA	1755.00		
FLOORING	SUBSTITUTE OTHER FLOOR COVERINGS FOR SPECIFIED SELECT OAK FLOORING				
	CARPET AND PAD @ $20/YD **DEDUCT**	SF	5.99		
	UNDERLAYMENT AND VINYL TILE OR SHEET VINYL **DEDUCT**	SF	5.54		
	SF = FLOOR				

GARAGE

DETACHED GARAGE	ATTACHED GARAGE
PREPARE PLANS AND OBTAIN PERMITS AS REQUIRED, PERMIT FEE COST **NOT** INCLUDED	PREPARE PLANS AND OBTAIN PERMITS AS REQUIRED, PERMIT FEE COST **NOT** INCLUDED
8" X 16" CONTINUOUS CONCRETE FOOTINGS, 36" BELOW GRADE WITH 8" X 8" X 16" BLOCK BUILDUP TO 12" ABOVE GRADE 4" CONCRETE FLOOR SLAB OVER GRAVEL	8" X 16" CONTINUOUS CONCRETE FOOTINGS, 36" BELOW GRADE WITH 8" X 8" X 16" BLOCK BUILDUP TO 12" ABOVE GRADE 4" CONCRETE FLOOR SLAB OVER GRAVEL WITH NO. 4 (1/2") REBAR INTO EXISTING HOUSE FOUNDATION WALL, 24" O.C.
2" X 4" STUD WALLS, 16" O.C. 1/2" PLYWOOD SHEATHING	2" X 4" STUD WALLS, 16" O.C. 1/2" PLYWOOD SHEATHING
GABLE TYPE ROOF WITH 12" OVERHANG, 4/12 SLOPE, CEILING JOISTS 1/2" PLYWOOD ROOF SHEATHING	GABLE TYPE ROOF WITH 12" OVERHANG, 4/12 SLOPE, CEILING JOISTS 1/2" PLYWOOD ROOF SHEATHING
#215 FIBERGLASS STRIP SHINGLES ALUMINUM GUTTERS AND DOWNSPOUTS AS REQUIRED	#215 FIBERGLASS STRIP SHINGLES ALUMINUM STEP FLASHING TO EXISTING HOUSE WALL ALUMINUM GUTTERS AND DOWNSPOUTS AS REQUIRED
TIGHT KNOT PINE OR FIR 6" FASCIA TIGHT KNOT PINE OR FIR 12" SOFFIT RAKE AND RAKE MOULDING	TIGHT KNOT PINE OR FIR 6" FASCIA TIGHT KNOT PINE OR FIR 12" SOFFIT RAKE AND RAKE MOULDING
8" HORIZONTAL VINYL SIDING, UNINSULATED	8" HORIZONTAL VINYL SIDING, UNINSULATED
ONE 8-0 X 7-0 FOUR-SECTION OVERHEAD GARAGE DOOR, INCLUDING EXTERIOR BRICK MOULDING, JAMBS, STOPS AND HARDWARE	ONE 8-0 X 7-0 FOUR-SECTION OVERHEAD GARAGE DOOR, INCLUDING EXTERIOR BRICK MOULDING, JAMBS, STOPS AND HARDWARE
TWO 2-8 X 4-2 WOOD DOUBLE HUNG WINDOWS, SINGLE GLAZED	TWO 2-8 X 4-2 WOOD DOUBLE HUNG WINDOWS, SINGLE GLAZED, INCLUDING INTERIOR CASINGS, STOOL AND STOOL CAP
	5/8" FIRECODE DRYWALL, TAPED AND FINISHED, ON EXISTING HOUSE WALL
REMOVE TRASH FROM PREMISES, DUMPING FEE **NOT** INCLUDED	REMOVE TRASH FROM PREMISES, DUMPING FEE **NOT** INCLUDED
NO PAINTING	**NO** PAINTING

SPECIFICATIONS		UNIT	PRICE	LOCAL AREA MODIFICATION %	LOCAL AREA MODIFICATION PRICE
DETACHED GARAGE	AS SPECIFIED	EA PLUS SF	7902.00 19.04		
ATTACHED GARAGE	AS SPECIFIED	EA PLUS SF	7902.00 18.47		

GARAGE EXTRAS AND ALLOWANCES

SPECIFICATIONS		UNIT	PRICE	LOCAL AREA MODIFICATION %	LOCAL AREA MODIFICATION PRICE
DIFFERENT FOOTING DEPTH	***Depth of Bottom of Footing Below Grade***				
	12" **DEDUCT**	SF	20.94		
	24" **DEDUCT**	SF	10.46		
	48" **ADD**	SF	10.43		
	LF = FOOTINGS				
ALTERNATE WALL SYSTEM	SUBSTITUTE MASONRY WALL FROM TOP OF SLAB TO START OF GARAGE ROOF				
	8" BLOCK **DEDUCT**	SF	1.14		
	BRICK & BLOCK **ADD**	SF	8.72		
	SF = SF GARAGE WALLS				
FURRING	FURR EXISTING HOUSE WALL WITH 1" X 3"				
	STRAIGHT WALL **ADD**	SF	.92		
	CROOKED WALL **ADD**	SF	1.16		
ROOF	OVERLAY EXISTING ROOF **ADD**	SF	12.33		
	SF = ROOF OVERLAY				
SIDING	REMOVE EXISTING WOOD, PLYWOOD OR ALUMINUM SIDING FROM EXISTING HOUSE WALL	SF	.51		
	SF = HOUSE WALL				
DOOR	SUBSTITUTE 16-0 X 7-0 OVERHEAD GARAGE DOOR **ADD**	EA	501.00		
	BREAK THROUGH EXISTING HOUSE WALL & INSTALL 3-0 X 6-8 X 3/4" FLUSH SOLID CORE DOOR TRIMMED BOTH SIDES				
	FRAME WALL **ADD**	EA	1194.00		
	BRICK VENEER **ADD**	EA	1251.00		
	BRICK AND BLOCK **ADD**	EA	1284.00		
	REMOVE EXISTING WINDOW AND WALL UNDER AND INSTALL 3-0 X 6-8 X 3/4" FLUSH SOLID CORE DOOR TRIMMED BOTH SIDES	EA	1050.00		
	INSTALL 3-0 X 6-8 THREE PANEL, 4-LIGHT DOOR, FULLY TRIMMED, FROM GARAGE TO OUTSIDE	EA	612.00		
WINDOWS	ADD OR OMIT ONE WOOD DOUBLE HUNG WINDOW, 2-8 X 4-2, SINGLE GLAZED **ADD** OR **DEDUCT**	EA	222.00		
ELECTRICAL	EXTEND ELECTRICAL CIRCUIT 50 LF TO DETACHED GARAGE **ADD**	EA	123.00		
	SWITCH, LIGHT AND GFIC OUTLET **ADD**	EA	240.00		
	EA = TOTAL AMOUNT				

SMALL KITCHENS

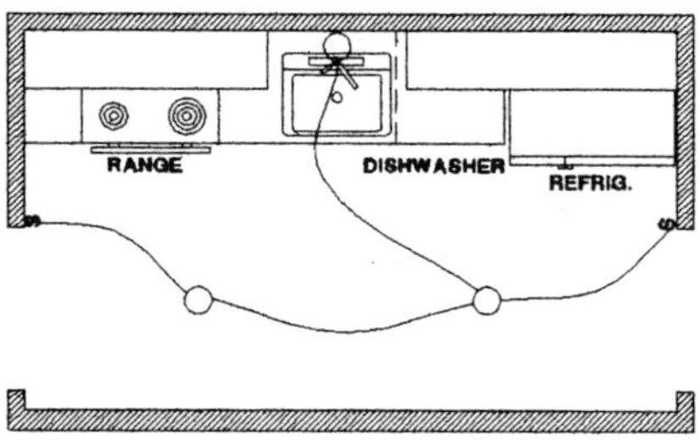

NO. 1 - 84 SF

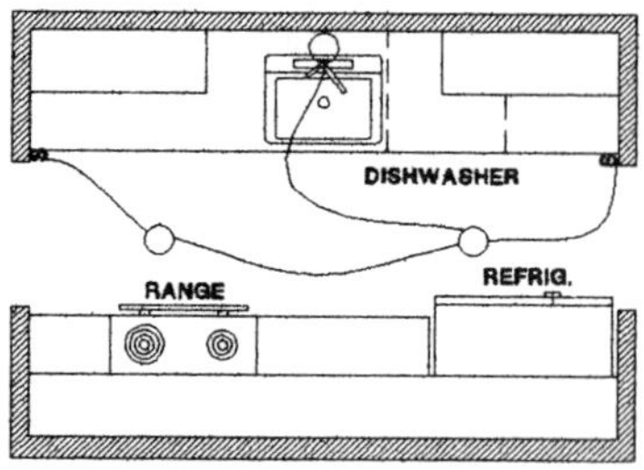

NO. 2 - 70 SF

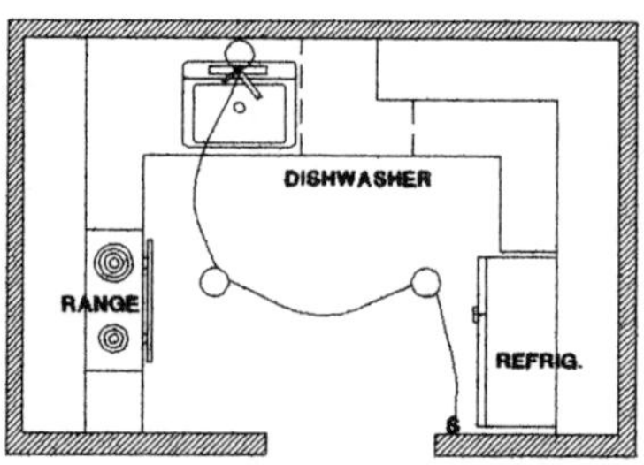

NO. 3 - 70 SF

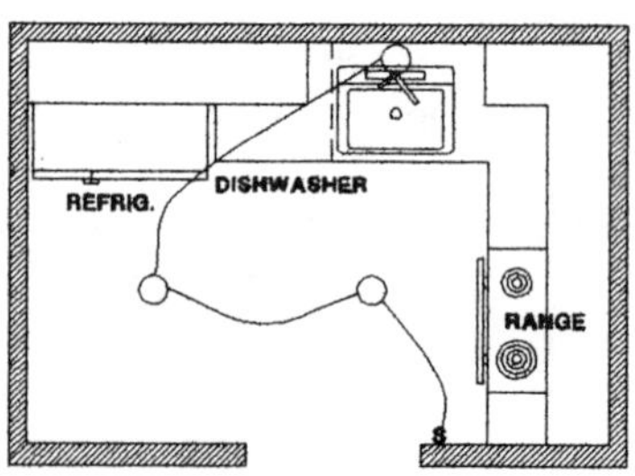

NO. 4 - 70 SF

REMODEL EXISTING KITCHEN	INSTALL NEW KITCHEN
PREPARE PLANS AND OBTAIN PERMIT, PERMIT FEE COST **NOT** INCLUDED	PREPARE PLANS AND OBTAIN PERMIT, PERMIT FEE COST **NOT** INCLUDED
TEAR OUT EXISTING KITCHEN CABINETS, COUNTERTOPS, SINK, FLOOR COVERING, STOVE, REFRIGERATOR & REMOVE TO OUTSIDE REPAIR MINOR DRYWALL DAMAGE	INSTALL 1/2" DRYWALL ON WALLS AND CEILING, TAPED AND FINISHED
FURNISH AND INSTALL WALL AND BASE CABINETS WITH DRAWER AND DOOR HARDWARE, INCLUDING 36" SINK BASE LAMINATE COUNTERTOPS WITH 4" BACKSPLASH AND SINK CUTOUT 24" X 21" SINGLE BOWL STAINLESS STEEL SINK WITH SINGLE LEVER FAUCET, FURNISHED AND INSTALLED IN EXISTING SINK LOCATION	FURNISH AND INSTALL WALL AND BASE CABINETS WITH DRAWER AND DOOR HARDWARE, INCLUDING 36" SINK BASE LAMINATE COUNTERTOPS WITH 4" BACKSPLASH AND SINK CUTOUT 24" X 21" SINGLE BOWL STAINLESS STEEL SINK WITH SINGLE LEVER FAUCET, FURNISHED AND INSTALLED WITHIN 5 FEET OF EXISTING STACK
REPLACE SWITCHES AND LIGHTS IN EXISTING LOCATIONS AS NECESSARY	SWITCHES AND LIGHTS AS SHOWN DUPLEX OUTLETS TO CODE APPLIANCE OUTLET 220-V RANGE OUTLET
FURNISH AND INSTALL ELECTRIC RANGE IN SAME LOCATION, 30" WHITE, BUILDER GRADE WITH DUCTLESS RANGE HOOD 1/2 HP DISPOSER ON EXISTING CIRCUIT AND PLUMBING DISHWASHER ON EXISTING CIRCUIT AND PLUMBING 16 CF REFRIGERATOR, DOUBLE DOOR, TOP FREEZER, FROST FREE, WHITE, BLDR. GRADE	FURNISH AND INSTALL ELECTRIC RANGE, INCLUDING DEDICATED CIRCUIT AS NECESSARY, 30" WHITE, BUILDER GRADE WITH DUCTLESS RANGE HOOD 1/2 HP DISPOSER ON NEW CIRCUIT AND PLUMBING DISHWASHER ON NEW CIRCUIT AND PLUMBING 16 CF REFRIGERATOR, DOUBLE DOOR, TOP FREEZER, FROST FREE, WHITE, BLDR. GRADE
INSTALL 3/8" PLYWOOD UNDERLAYMENT AND SHEET VINYL LAID IN ADHESIVE, INCLUDING 4" VINYL BASE AND CORNERS	INSTALL 3/8" PLYWOOD UNDERLAYMENT AND SHEET VINYL LAID IN ADHESIVE, INCLUDING 4" VINYL BASE AND CORNERS
PRIME ROOM AS NECESSARY AND PAINT WITH 2 COATS TOP QUALITY SEMI-GLOSS ENAMEL	PRIME ENTIRE ROOM AND PAINT WITH 2 COATS TOP QUALITY SEMI-GLOSS ENAMEL
REMOVE OLD CABINETS, FIXTURES, APPLIANCES & TRASH FROM PREMISES & FINAL CLEAN-UP, DUMPING FEE **NOT** INCLUDED	REMOVE TRASH FROM PREMISES AND FINAL CLEAN-UP, DUMPING FEE **NOT** INCLUDED

SPECIFICATIONS			UNIT	PRICE	LOCAL AREA MODIFICATION	
					%	PRICE
REMODEL KITCHEN	REMODEL BUILDER QUALITY KITCHEN	NO. 1	EA	7638.00		
		NO. 2	EA	10,206.00		
		NO. 3	EA	10,335.00		
		NO. 4	EA	8964.00		
INSTALL NEW KITCHEN	INSTALL NEW BUILDER QUALITY KITCHEN	NO. 1	EA	9429.00		
		NO. 2	EA	11,928.00		
		NO. 3	EA	11,976.00		
		NO. 4	EA	10,662.00		

LARGE KITCHENS

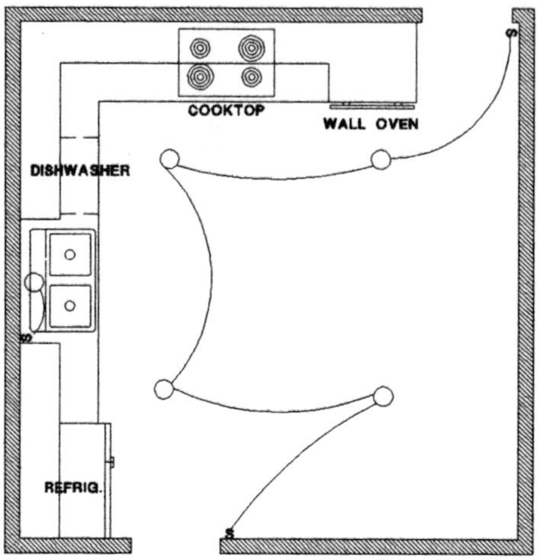

NO. 1 - 175 SF

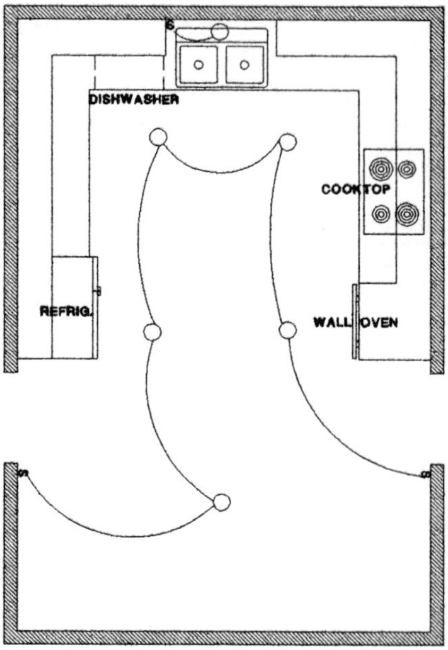

NO. 2 - 216 SF

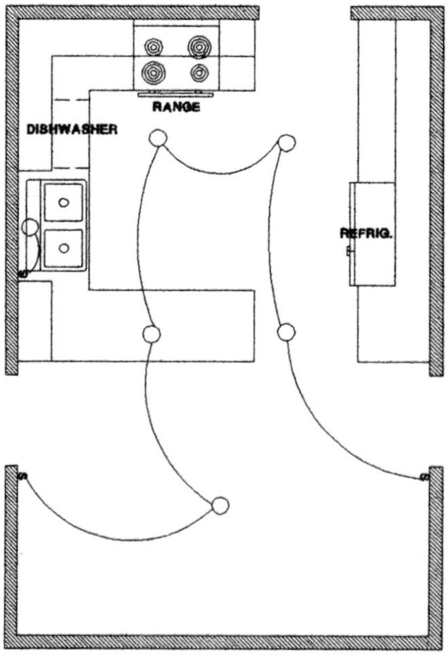

NO. 3 - 216 SF

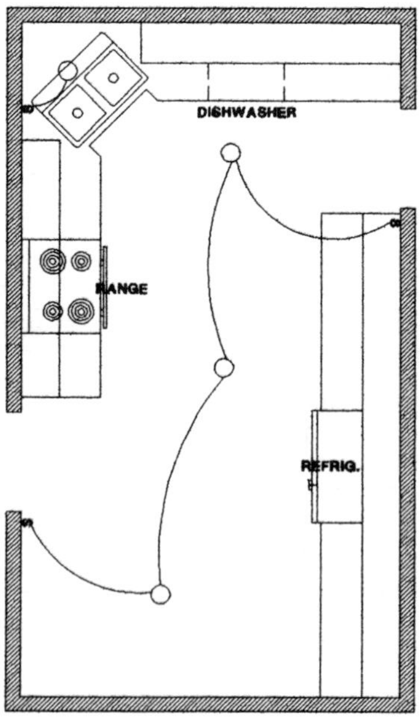

NO. 4 - 180 SF

REMODEL EXISTING KITCHEN	INSTALL NEW KITCHEN
PREPARE PLANS AND OBTAIN PERMIT, PERMIT FEE COST **NOT** INCLUDED	PREPARE PLANS AND OBTAIN PERMIT, PERMIT FEE COST **NOT** INCLUDED
TEAR OUT EXISTING KITCHEN CABINETS, COUNTERTOPS, SINK, FLOOR COVERING, STOVE AND REFRIGERATOR AND REMOVE TO OUTSIDE REPAIR MINOR DRYWALL DAMAGE	INSTALL 1/2" DRYWALL ON WALLS AND CEILING, TAPED AND FINISHED
FURNISH AND INSTALL WALL AND BASE CABINETS WITH DRAWER AND DOOR HARDWARE, INCLUDING 36" SINK BASE LAMINATE COUNTERTOPS WITH 4" BACKSPLASH AND SINK CUTOUT 24" X 21" SINGLE BOWL STAINLESS STEEL SINK WITH SINGLE LEVER FAUCET, FURNISHED AND INSTALLED IN EXISTING SINK LOCATION	FURNISH AND INSTALL WALL AND BASE CABINETS WITH DRAWER AND DOOR HARDWARE, INCLUDING 36" SINK BASE LAMINATE COUNTERTOPS WITH 4" BACKSPLASH AND SINK CUTOUT 24" X 21" SINGLE BOWL STAINLESS STEEL SINK WITH SINGLE LEVER FAUCET, FURNISHED AND INSTALLED WITHIN 5 FEET OF EXISTING STACK
REPLACE SWITCHES AND LIGHTS IN EXISTING LOCATIONS AS NECESSARY	SWITCHES AND LIGHTS AS SHOWN DUPLEX OUTLETS TO CODE APPLIANCE OUTLET 220-V RANGE OUTLET
FURNISH AND INSTALL ELECTRIC RANGE IN SAME LOCATION, 30" WHITE, BUILDER GRADE WITH DUCTLESS RANGE HOOD 1/2 HP DISPOSER ON EXISTING CIRCUIT AND PLUMBING DISHWASHER ON EXISTING CIRCUIT AND PLUMBING 16 CF REFRIGERATOR, DOUBLE DOOR, TOP FREEZER, FROST FREE, WHITE, BLDR. GRADE	FURNISH AND INSTALL ELECTRIC RANGE, INCLUDING DEDICATED CIRCUIT AS NECESSARY, 30" WHITE, BUILDER GRADE WITH DUCTLESS RANGE HOOD 1/2 HP DISPOSER ON NEW CIRCUIT AND PLUMBING DISHWASHER ON NEW CIRCUIT AND PLUMBING 16 CF REFRIGERATOR, DOUBLE DOOR, TOP FREEZER, FROST FREE, WHITE, BLDR. GRADE
INSTALL 3/8" PLYWOOD UNDERLAYMENT AND SHEET VINYL LAID IN ADHESIVE, INCLUDING 4" VINYL BASE AND CORNERS	INSTALL 3/8" PLYWOOD UNDERLAYMENT AND SHEET VINYL LAID IN ADHESIVE, INCLUDING 4" VINYL BASE AND CORNERS
PRIME ROOM AS NECESSARY AND PAINT WITH 2 COATS TOP QUALITY SEMI-GLOSS ENAMEL	PRIME ENTIRE ROOM AND PAINT WITH 2 COATS TOP QUALITY SEMI-GLOSS ENAMEL
REMOVE OLD CABINETS, FIXTURES, APPLIANCES AND TRASH FROM PREMISES AND FINAL CLEAN-UP, DUMPING FEE **NOT** INCLUDED	REMOVE TRASH FROM PREMISES AND FINAL CLEAN-UP, DUMPING FEE **NOT** INCLUDED

SPECIFICATIONS			UNIT	PRICE	LOCAL AREA MODIFICATION	
					%	PRICE
REMODEL KITCHEN	REMODEL BUILDER QUALITY KITCHEN	NO. 1	EA	13,401.00		
		NO. 2	EA	16,164.00		
		NO. 3	EA	13,437.00		
		NO. 4	EA	14,316.00		
INSTALL NEW KITCHEN	INSTALL NEW BUILDER QUALITY KITCHEN	NO. 1	EA	13,254.00		
		NO. 2	EA	16,497.00		
		NO. 3	EA	13,590.00		
		NO. 4	EA	14,877.00		

ISLAND KITCHENS

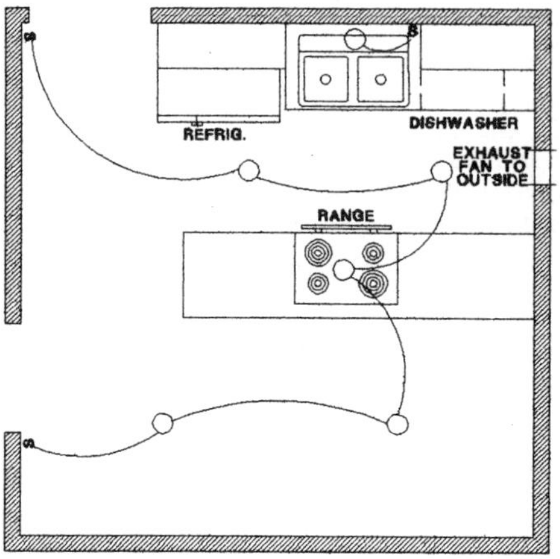

NO. 1 - 156 SF

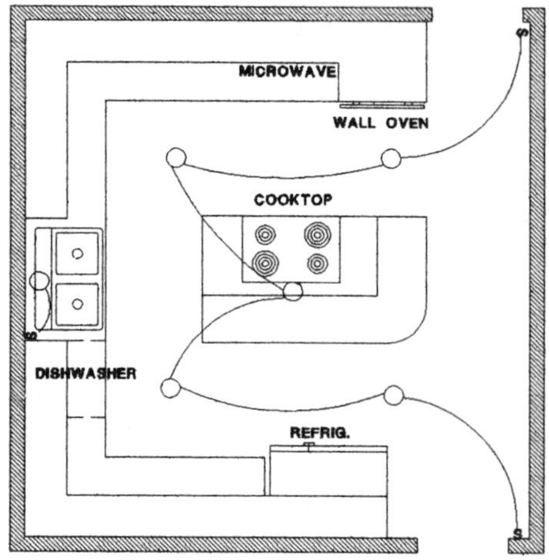

NO. 2 - 175 SF

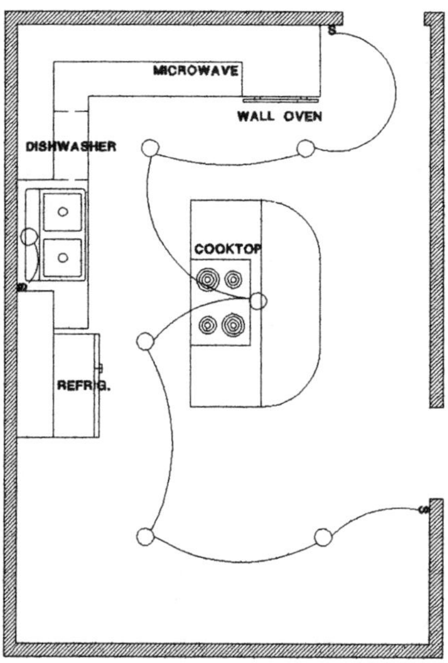

NO. 3 - 216 SF

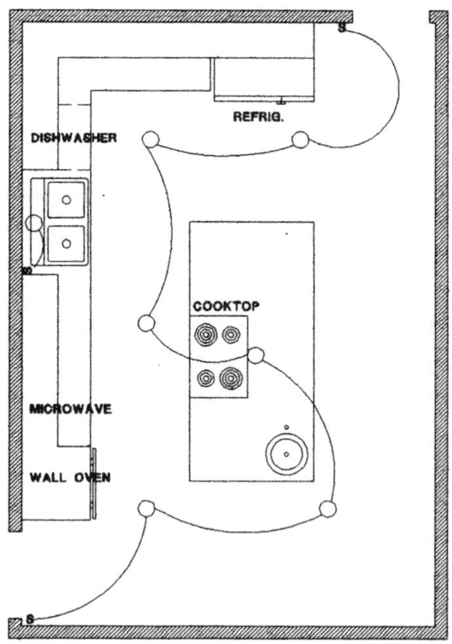

NO. 4 - 231 SF

REMODEL EXISTING KITCHEN	INSTALL NEW KITCHEN
PREPARE PLANS AND OBTAIN PERMIT, PERMIT FEE COST **NOT** INCLUDED	PREPARE PLANS AND OBTAIN PERMIT, PERMIT FEE COST **NOT** INCLUDED
TEAR OUT EXISTING KITCHEN CABINETS, COUNTERTOPS, SINK, FLOOR COVERING, STOVE AND REFRIGERATOR AND REMOVE TO OUTSIDE REPAIR MINOR DRYWALL DAMAGE	INSTALL 1/2" DRYWALL ON WALLS AND CEILING, TAPED AND FINISHED
FURNISH AND INSTALL WALL AND BASE CABINETS, INCLUDING ISLAND BASE AND 36" SINK BASE WITH DRAWER AND DOOR HARDWARE LAMINATE COUNTERTOPS WITH 4" BACKSPLASH AND SINK CUTOUTS 32" X 21" DOUBLE BOWL STAINLESS STEEL SINK WITH SINGLE LEVER FAUCET 15" X 15" BAR SINK WITH BAR FAUCET (KITCHEN NO. 4) FURNISHED AND INSTALLED IN EXISTING SINK LOCATION	FURNISH AND INSTALL WALL AND BASE CABINETS, INCLUDING ISLAND BASE AND 36" SINK BASE WITH DRAWER AND DOOR HARDWARE LAMINATE COUNTERTOPS WITH 4" BACKSPLASH AND SINK CUTOUTS 32" X 21" DOUBLE BOWL STAINLESS STEEL SINK WITH SINGLE LEVER FAUCET FURNISHED & INSTALLED WITHIN 5 FT. OF EXISTING STACK 15" X 15" BAR SINK WITH BAR FAUCET (KITCHEN NO. 4) FURNISHED AND INSTALLED WITHIN 5 FEET OF EXISTING STACK
REPLACE SWITCHES AND LIGHTS IN EXISTING LOCATIONS AS NECESSARY	SWITCHES AND LIGHTS AS SHOWN DUPLEX OUTLETS TO CODE APPLIANCE OUTLET 220-V RANGE OUTLET
FURNISH AND INSTALL ELECTRIC COOKTOP WITH DOWNDRAFT DUCTED EXHAUST AND ELECTRIC WALL OVEN IN SAME LOCATION 30" BUILT-IN MICROWAVE OVEN 1/2 HP DISPOSER ON EXISTING CIRCUIT AND PLUMBING DISHWASHER ON EXISTING CIRCUIT AND PLUMBING 16 CF REFRIGERATOR, DOUBLE DOOR, TOP FREEZER, FROST FREE, WHITE, BLDR. GRADE	FURNISH AND INSTALL ELECTRIC COOKTOP WITH DOWNDRAFT DUCTED EXHAUST AND ELECTRIC WALL OVEN, INCLUDING DEDICATED CIRCUIT AS NECESSARY 30" BUILT-IN MICROWAVE OVEN 1/2 HP DISPOSER ON NEW CIRCUIT AND PLUMBING DISHWASHER ON NEW CIRCUIT AND PLUMBING 16 CF REFRIGERATOR, DOUBLE DOOR, TOP FREEZER, FROST FREE, WHITE, BLDR. GRADE
INSTALL 3/8" PLYWOOD UNDERLAYMENT AND SHEET VINYL LAID IN ADHESIVE, INCLUDING 4" VINYL BASE AND CORNERS	INSTALL 3/8" PLYWOOD UNDERLAYMENT AND SHEET VINYL LAID IN ADHESIVE, INCLUDING 4" VINYL BASE AND CORNERS
PRIME ROOM AS NECESSARY AND PAINT WITH 2 COATS TOP QUALITY SEMI-GLOSS ENAMEL	PRIME ENTIRE ROOM AND PAINT WITH 2 COATS TOP QUALITY SEMI-GLOSS ENAMEL
REMOVE OLD CABINETS, FIXTURES, APPLIANCES AND TRASH FROM PREMISES AND FINAL CLEAN-UP, DUMPING FEE **NOT** INCLUDED	REMOVE TRASH FROM PREMISES AND FINAL CLEAN-UP, DUMPING FEE **NOT** INCLUDED

SPECIFICATIONS			UNIT	PRICE	LOCAL AREA MODIFICATION %	LOCAL AREA MODIFICATION PRICE
REMODEL KITCHEN	REMODEL BUILDER QUALITY KITCHEN	NO. 1	EA	11,451.00		
		NO. 2	EA	18,378.00		
		NO. 3	EA	15,258.00		
		NO. 4	EA	16,530.00		
INSTALL NEW KITCHEN	INSTALL NEW BUILDER QUALITY KITCHEN	NO. 1	EA	12,081.00		
		NO. 2	EA	19,665.00		
		NO. 3	EA	16,629.00		
		NO. 4	EA	18,414.00		

KITCHEN — EXTRAS & ALLOWANCES

SPECIFICATIONS			UNIT	PRICE	LOCAL AREA MODIFICATION %	LOCAL AREA MODIFICATION PRICE
ADDITIONAL AREA	ADDITIONAL FLOOR, WALL AND CEILING AREA FOR KITCHEN (ADDITIONAL CABINETS AND APPLIANCES MAY REQUIRE ADDITIONAL FLOOR SPACE) SF = FLOOR AREA OF ADDITIONAL SPACE		SF	15.80		
SUBSTITUTE CABINET QUALITY	SUBSTITUTE ECONOMY QUALITY CABINETS					
	SMALL KITCHEN 1	**DEDUCT**	EA	603.00		
	2	**DEDUCT**	EA	1314.00		
	3	**DEDUCT**	EA	1326.00		
	4	**DEDUCT**	EA	951.00		
	LARGE KITCHEN 1	**DEDUCT**	EA	1446.00		
	2	**DEDUCT**	EA	2172.00		
	3	**DEDUCT**	EA	1515.00		
	4	**DEDUCT**	EA	1941.00		
	ISLAND KITCHEN 1	**DEDUCT**	EA	894.00		
	2	**DEDUCT**	EA	2298.00		
	3	**DEDUCT**	EA	1557.00		
	4	**DEDUCT**	EA	1692.00		
	SUBSTITUTE PREMIUM QUALITY CABINETS					
	SMALL KITCHEN 1	**ADD**	EA	576.00		
	2	**ADD**	EA	1284.00		
	3	**ADD**	EA	1287.00		
	4	**ADD**	EA	918.00		
	LARGE KITCHEN 1	**ADD**	EA	1434.00		
	2	**ADD**	EA	1719.00		
	3	**ADD**	EA	1482.00		
	4	**ADD**	EA	1866.00		
	ISLAND KITCHEN 1	**ADD**	EA	906.00		
	2	**ADD**	EA	2397.00		
	3	**ADD**	EA	1584.00		
	4	**ADD**	EA	2247.00		
ADD OR OMIT CABINETS	ADD OR OMIT BASE CABINETS					
	ECONOMY	**ADD** OR **DEDUCT**	LF	114.00		
	BUILDER	**ADD** OR **DEDUCT**	LF	165.00		
	PREMIUM	**ADD** OR **DEDUCT**	LF	222.00		
	LF = FRONT OF CABINETS					
	ADD OR OMIT WALL CABINETS					
	ECONOMY	**ADD** OR **DEDUCT**	LF	78.00		
	BUILDER	**ADD** OR **DEDUCT**	LF	114.00		
	PREMIUM	**ADD** OR **DEDUCT**	LF	144.00		
	LF = FRONT OF CABINETS					
BROOM CLOSET	SUBSTITUTE 18" X 24" BROOM CLOSET FOR BASE AND WALL CABINETS					
	ECONOMY	**ADD**	EA	45.00		
	BUILDER	**ADD**	EA	69.00		
	PREMIUM	**ADD**	EA	123.00		
	ADD 18" X 24" BROOM CLOSET	**ADD**	EA	483.00		

SPECIFICATIONS			UNIT	PRICE	LOCAL AREA MODIFICATION	
					%	PRICE
COUNTER-TOP	SUBSTITUTE COUNTERTOP WITH 4" BACKSPLASH					
	CORIAN	**ADD**	LF	117.00		
	BUTCHER BLOCK	**ADD**	LF	63.00		
	CERAMIC TILE	**ADD**	LF	18.00		
BULKHEAD	DRYWALL BULKHEAD FROM CEILING TO WALL CABINETS, MOULDING AT JOINT	**ADD**	LF	16.95		
PLUMBING	SUBSTITUTE 33" X 21" CAST IRON PORCELAIN FINISHED DOUBLE BOWL SINK FOR SPECIFIED STAINLESS STEEL SINGLE BOWL	**ADD**	EA	15.00		
	ADD ICEMAKER TO FACTORY EQUIPPED REFRIGERATOR	**ADD**	EA	150.00		
	INSTALL INSTANT HOT AT SINK, INCLUDING ELECTRICAL HOOKUP	**ADD**	EA	177.00		
	EXTEND EXISTING PLUMBING AND ROUGH IN FOR KITCHEN SINK, OVER FIVE FEET FROM EXISTING SINK LF = TOTAL DISTANCE FROM EXISTING SINK		LF	42.00		
ELECTRICAL	ADDITIONAL LIGHT AND SWITCH (DOES **NOT** INCLUDE COST OF FIXTURE)	**ADD**	EA	111.00		
	ADDITIONAL DUPLEX OUTLET	**ADD**	EA	57.00		
	SUBSTITUTE DIMMER SWITCH FOR SPECIFIED LIGHT AND SWITCH	**ADD**	EA	21.00		
FLOOR	SUBSTITUTE KITCHEN FLOOR COVERING					
	CERAMIC TILE	**ADD**	SF	17.91		
	12" X 12" MARBLE	**ADD**	SF	21.30		
	T&G SELECT OAK FLOORING, SANDED AND FINISHED	**ADD**	SF	5.51		
	12" X 12" SLATE	**ADD**	SF	7.53		

OPEN PORCH

SPECIFICATIONS	
PLANS AND PERMIT	PREPARE PLANS AND OBTAIN PERMITS AS REQUIRED, PERMIT FEE COST **NOT** INCLUDED
CONCRETE AND MASONRY	EXCAVATE TO 36" BELOW GRADE, FORM AND POUR CONTINUOUS CONCRETE FOOTINGS WITH REBAR BLOCK FOUNDATION WALL ON FOOTINGS TO 18" ABOVE GRADE, BACKFILL AND ROUGH GRADE AS REQUIRED 4" THICK CONCRETE FLOOR SLAB OVER GRAVEL BASE
WALL FRAMING	4" X 4" PRESSURE TREATED FIR OR PINE POSTS, 6'-0" O.C. 4" X 10" HEADERS ON POSTS AT PORCH PERIMETER 2" X 4" STUDS ON GABLE END WITH 1/2" PLYWOOD SHEATHING
ROOF FRAMING	GABLE TYPE ROOF WITH 12" OVERHANG, RAFTERS AND CEILING JOISTS AS REQUIRED 1/2" CDX PLYWOOD ROOF SHEATHING
ROOF COVERING, GUTTERS, FLASHING	#215 FIBERGLASS SHINGLES OVER #15 FELT PAPER ALUMINUM GUTTERS AND DOWNSPOUTS ALUMINUM FLASHING WHERE ROOF JOINS EXISTING HOUSE
EXTERIOR TRIM	TRIM HEADERS THREE SIDES WITH TIGHT KNOT PINE OR FIR TIGHT KNOT PINE OR FIR FASCIA AND SOFFIT RAKE AND RAKE MOULDING ON GABLE END
SIDING	VINYL SIDING ON GABLE END
CEILING	FINISH PORCH CEILING WITH 5/8" X 4" FIR BEADED CEILING WITH 3/4" CEILING COVE MOULDING AT HEADER INTERSECTIONS
CLEAN-UP	REMOVE TRASH FROM PREMISES, DUMPING FEE **NOT** INCLUDED
	NO PAINTING

SPECIFICATIONS		UNIT	PRICE	LOCAL AREA MODIFICATION	
				%	PRICE
OPEN PORCH	AS SPECIFIED	EA PLUS	3825.00		
	SF = PORCH FLOOR	SF	17.25		

OPEN PORCH — EXTRAS & ALLOWANCES

SPECIFICATIONS		UNIT	PRICE	LOCAL AREA MODIFICATION %	LOCAL AREA MODIFICATION PRICE
DIFFERENT FOOTING DEPTH	***Depth of Bottom of Footing Below Grade***				
	12" **DEDUCT**	LF	20.94		
	24" **DEDUCT**	LF	10.46		
	48" **DEDUCT**	LF	10.43		
	LF = FOOTINGS				
SUBSTITUTE FLOOR SYSTEM	SUBSTITUTE MUDSILL, FLOOR JOISTS AND 3/4" T & G FIR OR PINE PORCH FLOORING FOR SPECIFIED CONCRETE **ADD**	SF	5.33		
	SF = PORCH FLOOR				
DOOR	REMOVE EXISTING WINDOW AND WALL UNDER AND INSTALL 2-8 X 6-8 THREE PANEL, 4 LIGHTS 1-3/4" DOOR, FULLY TRIMMED, FROM HOUSE TO PORCH	EA	1239.00		
	BREAK THROUGH EXISTING HOUSE WALL AND INSTALL 2-8 X 6-8 THREE PANEL, 4 LIGHTS 1-3/4" DOOR, FULLY TRIMMED, FROM HOUSE TO PORCH				
	FRAME WALL **ADD**	EA	1380.00		
	BRICK & FRAME **ADD**	EA	1437.00		
	BRICK & BLOCK **ADD**	EA	1470.00		
SCREEN	ALUMINUM SCREENING ON EXISTING POSTS WITH PANEL STRIPS **ADD**	SF	4.14		
SCREEN DOOR, WOOD	WOOD FRAME SCREEN DOOR WITH ALUMINUM SCREENING, INCLUDING ALL NECESSARY HARDWARE FOR INSTALLATION **ADD**	EA	168.00		

BASEMENT RECREATION ROOM

SPECIFICATIONS	
PLANS AND PERMIT	PREPARE PLANS AND OBTAIN PERMITS AS REQUIRED, PERMIT FEE COST **NOT** INCLUDED
WALL FRAMING	1" X 3" FURRING, 16" O.C., AND/OR 2" X 3" STUDDING, 16" O.C., ON FOUR WALLS
DOOR	ONE 2-6 X 6-8 X 1-3/8" HOLLOW CORE FLUSH DOOR, INCLUDING HARDWARE AND PREFINISHED TRIM BOTH SIDES
ELECTRICAL	DUPLEX OUTLETS TO NATIONAL CODE ONE RECESSED CEILING FIXTURE AND SWITCH
WALLS	1/4" PREFINISHED PANELING ON FOUR WALLS @ $20.00 PER SHEET
TRIM	ON PANELING: 3-1/2" PREFINISHED CLAM BASE AND 1-5/8" PREFINISHED CEILING MOULDING TRIM ONE EXISTING WINDOW WITH PLYWOOD OR DRYWALL AND WOOD CORNER GUARD
CEILING	1" X 3" FURRING ACROSS JOISTS, 12" O.C., AND 12" X 24" PLAIN WHITE CEILING TILE @ .40 PSF RETAIL
CLEAN-UP	REMOVE TRASH FROM PREMISES, DUMPING FEE **NOT** INCLUDED
	NO PAINTING

SPECIFICATIONS		UNIT	PRICE	LOCAL AREA MODIFICATION	
				%	PRICE
RECREATION ROOM	AS SPECIFIED	EA	2364.00		
		PLUS			
	SF = FLOOR AREA	SF	11.13		

RECREATION ROOM — EXTRAS & ALLOWANCES

SPECIFICATIONS		UNIT	PRICE	LOCAL AREA MODIFICATION	
				%	PRICE
INTERIOR PARTITION	BUILD COMPLETE PARTITION WITH FOLLOWING SPECIFICATIONS: • WALL FRAMING, 2 X 4 STUDS AND PLATES, 8 FOOT HIGH WALL • DUPLEX OUTLETS TO CODE BOTH SIDES OF WALL • 1/4" PREFINISHED PLYWOOD PANELING @ $15 PER SHEET BOTH SIDES OF WALL • 3-1/2" BASE AND SHOE BOTH SIDES OF WALL LF = LENGTH OF PARTITION	LF	60.00		
TEAR OUT OLD FLOOR, INSTALL NEW FLOOR	BREAK UP EXISTING FLOOR SLAB IN BASEMENT AND HAUL RUBBLE FROM PREMISES, INSTALL NEW BASEMENT FLOOR, 4" CONCRETE WITH PLASTIC VAPOR BARRIER UNDER	SF	8.87		
HEAT	INSTALL REGISTER IN EXISTING HEAT DUCT	EA	180.00		
INSULATION	INSULATE EXTERIOR WALLS AND/OR CEILING				
	3/4" FOIL FACED FOAM	SF	.59		
	3-1/2" BLANKET	SF	.72		

CONVERSION — ATTACHED GARAGE TO ROOM

SPECIFICATIONS	
PLANS AND PERMIT	PREPARE PLANS AND OBTAIN PERMITS AS REQUIRED, PERMIT FEE COST **NOT** INCLUDED
TEAR-OUT	REMOVE EXISTING OVERHEAD GARAGE DOOR AND TRIM
FLOOR FRAMING	BUILD UP A LEVEL FLOOR ABOVE EXISTING CONCRETE WITH 2" X 4", 2" X 6" OR 2" X 8" SLEEPERS, 16" O.C.
	3/4" T & G FIR PLYWOOD SUBFLOOR, GLUED AND NAILED
WALL FRAMING	FRAME IN FOR THREE WINDOWS IN EXISTING GARAGE DOOR OPENING WITH 2" X 4" STUDS AND 1/2" CDX PLYWOOD SHEATHING
	FURR EXISTING HOUSE WALL WITH 1" X 3" FURRING
SIDING	8" HORIZONTAL VINYL SIDING, INSULATED, BELOW WINDOW OPENING
WINDOWS	THREE WOOD DOUBLE HUNG, DOUBLE GLAZED WINDOWS AND FRAMES, FULLY TRIMMED
	TRIM TWO EXISTING WINDOWS WITH CASINGS, STOOLS AND STOOL CAPS
HEATING	EXTEND HEAT FROM MAIN SYSTEM WITH WARM AIR HEAT DUCTS OR ELECTRIC BASEBOARD
ELECTRICAL	DUPLEX WALL OUTLETS TO CODE ON EXISTING SERVICE
INSULATION	3-1/2" FIBERGLASS BLANKET INSULATION IN EXTERIOR WALLS
	6" FIBERGLASS BLANKET INSULATION IN FLOOR AND CEILING
INTERIOR WALLS	1/2" DRYWALL ON FOUR WALLS, TAPED AND FINISHED READY FOR PAINTING
CEILING COVERING	1/2" DRYWALL, TAPED AND FINISHED READY FOR PAINTING
INTERIOR TRIM	3-1/2" BASE AND SHOE MOULDING ON ALL FOUR WALLS
FLOOR COVERING	2-1/4" SELECT OAK FLOORING, SANDED AND FINISHED
CLEAN-UP	REMOVE TRASH FROM PREMISES, DUMPING FEE **NOT** INCLUDED
	NO PAINTING

SPECIFICATIONS		UNIT	PRICE	LOCAL AREA MODIFICATION	
				%	PRICE
CONVERT GARAGE TO ROOM	AS SPECIFIED ABOVE	EA PLUS	3360.00		
	SF = SQUARE FOOT FLOOR AREA OF NEW ROOM	SF	21.71		
	ADD OR OMIT TRIMMING ONE WOOD DOUBLE HUNG WINDOW **ADD** OR **DEDUCT**	EA	75.00		
	INSTALL ADDITIONAL VINYL SIDING ON FRONT, INCLUDING REMOVING EXISTING SIDING **ADD** SF = ADDITIONAL SIDING	SF	4.05		
	UNDERPIN EXISTING SLAB AND INTERMEDIATE POINTS WITH 24" X 24" CONCRETE FOOTINGS 48" BELOW GRADE **ADD**	EA	639.00		

CONVERSION — OPEN PORCH TO ROOM

SPECIFICATIONS	
PLANS AND PERMIT	PREPARE PLANS AND OBTAIN PERMITS AS REQUIRED, PERMIT FEE COST **NOT** INCLUDED
TEAR-OUT	REMOVE EXISTING CORNER, END AND INTERMEDIATE POSTS, PORCH SCREENING AND CEILING COVERING INSTALL TEMPORARY SUPPORT FOR ROOF STRUCTURE
FLOOR FRAMING	BUILD UP A LEVEL FLOOR ABOVE EXISTING CONCRETE OR WOOD DECK WITH 2" X 4", 2" X 6" OR 2" X 8" SLEEPERS, 16" O.C. 3/4" T & G FIR PLYWOOD SUBFLOOR, GLUED AND NAILED
WALL FRAMING	2" X 4" STUDS, 16" O.C., AND 1/2" CDX PLYWOOD SHEATHING AT PORCH PERIMETER FURR EXISTING HOUSE WALL WITH 1" X 3" FURRING
SIDING	8" HORIZONTAL VINYL SIDING, INSULATED
WINDOWS	ONE 2-8 X 4-6 WOOD DOUBLE HUNG, DOUBLE GLAZED WINDOW FOR EACH 100 SF OF LIVING AREA, FULLY TRIMMED
HEATING	EXTEND HEAT FROM MAIN SYSTEM WITH WARM AIR HEAT DUCTS OR ELECTRIC BASEBOARD
ELECTRICAL	DUPLEX WALL OUTLETS TO CODE ON EXISTING SERVICE
INSULATION	3-1/2" FIBERGLASS BLANKET INSULATION IN EXTERIOR WALLS 6" FIBERGLASS BLANKET INSULATION IN FLOOR AND CEILING
INTERIOR WALL COVERING	1/2" DRYWALL ON WALLS, TAPED AND FINISHED READY FOR PAINTING
CEILING COVERING	1/2" DRYWALL, TAPED AND FINISHED READY FOR PAINTING
INTERIOR TRIM	3-1/2" BASE AND SHOE MOULDING ON ALL FOUR WALLS
FLOOR COVERING	2-1/4" SELECT OAK FLOORING, SANDED AND FINISHED
CLEAN-UP	REMOVE TRASH FROM PREMISES, DUMPING FEE **NOT** INCLUDED
	NO PAINTING

SPECIFICATIONS		UNIT	PRICE	LOCAL AREA MODIFICATION	
				%	PRICE
CONVERT OPEN PORCH TO ROOM	AS SPECIFIED ABOVE	EA	2760.00		
		PLUS			
	SF = SQUARE FOOT FLOOR AREA OF NEW ROOM	SF	29.40		
EXTRAS & ALLOW-ANCES	ADD OR OMIT ONE WOOD DOUBLE HUNG WINDOW **ADD** OR **DEDUCT**	EA	378.00		
	UNDERPIN EXISTING CONCRETE SLAB AND INTER-MEDIATE POINTS WITH 24" X 24" CONCRETE FOOTINGS 48" BELOW GRADE **ADD**	EA	639.00		
	OMIT REMOVAL OF PORCH FLOORING **DEDUCT** SF = FLOORING	SF	.75		

CONVERSION — SINGLE APARTMENT

SPECIFICATIONS	
	CONVERT SINGLE FLOOR (OR BASEMENT) OF EXISTING HOUSE TO SINGLE APARTMENT AS SPECIFIED BELOW:
PLANS & PERMIT	PREPARE PLANS AND OBTAIN PERMITS AS REQUIRED, PERMIT FEE COST **NOT** INCLUDED
TEAR-OUT	REMOVE EXISTING NON-BEARING WALLS REMOVE WALL COVERING FROM BEARING WALLS
FRAMING	FRAME INTERIOR PARTITIONS ACCORDING TO PLAN FURR EXTERIOR WALLS WITH 1" X 3" FURRING STRIPS
DOORS	NEW PAINT GRADE INTERIOR FLUSH DOORS, 1-3/8" TRIM DOORS WITH RANCH CASING
WINDOWS	RE-TRIM WINDOWS WITH RANCH CASING, STOP, STOOL AND APRON
PLUMBING	FULL BATH WITH LAVATORY, TUB AND W.C., PLUMBING FOR KITCHEN SINK AND DISPOSAL
HEATING & AIR CONDITIONING	SEPARATE FORCED AIR HEAT AND COOLING SYSTEM, ELECTRIC OR GAS, INCLUDING DUCTWORK AND REGISTERS
ELECTRICAL	150-AMP SERVICE WITH CIRCUIT BREAKERS, WIRE APARTMENT TO CODE, INCLUDING WALL OUTLETS, SWITCHES, FIXTURES, KITCHEN APPLIANCE OUTLETS
INSULATION	INSULATE EXTERIOR WALL WITH ALUMINUM FOIL ON FURRING STRIPS
INTERIOR WALLS AND CEILINGS	1/2" DRYWALL ON ALL WALLS, 5/8" DRYWALL ON ALL CEILINGS: TAPE AND FINISH 3 COATS AND SAND, READY FOR PAINTING CERAMIC TILE WITH MASTIC 6 FEET ABOVE FLOOR IN BATHTUB AREA ONLY
MILLWORK, TRIM	RANCH BASE AND OAK SHOE THROUGHOUT LINEN CLOSET 24" DEEP, 36" WIDE WITH FULL DEPTH SHELVES CLOSET TRIM IN BEDROOM CLOSET
CABINETS AND APPLIANCES	10 LF OF UPPER AND LOWER KITCHEN CABINETS (RANGE, SINK AND REFRIGERATOR AREAS INCLUDED IN MEASUREMENT) COUNTERTOP WITH STAINLESS STEEL SINK AND DISPOSAL NEW GAS OR ELECTRIC RANGE NEW 30-GALLON GAS OR ELECTRIC HOT WATER HEATER DUCT TYPE KITCHEN FAN, HOOD AND VENT TO OUTSIDE BATHROOM FAN, VENTED TO OUTSIDE
SPECIALTIES	HINGED DOOR MEDICINE CABINET WITH TWO FLUORESCENT LIGHTS PAPER HOLDER AND TOWEL BAR IN BATHROOM
FLOOR	VINYL COMPOSITION TILE IN KITCHEN, CERAMIC TILE IN BATHROOM, AND $12/YD CARPET WITH PAD IN ALL OTHER ROOMS
PAINTING	PAINT ALL WALLS AND CEILINGS, 2 COATS FLAT PAINT PAINT ALL TRIM, BATH AND KITCHEN, WITH 2 COATS SEMI-GLOSS
CLEAN-UP	CLEAN UP ALL RUBBISH DURING AND AT COMPLETION OF JOB AND REMOVE TO DUMPING GROUND WITHIN FIVE MILES DUMPSTER AND DUMPING FEE **NOT** INCLUDED

SPECIFICATIONS		UNIT	PRICE	LOCAL AREA MODIFICATION %	LOCAL AREA MODIFICATION PRICE
	NOTE: SF = SQUARE FOOT LIVING AREA, INCLUDING UTILITY ROOM				
SINGLE APARTMENT AS SPECIFIED	EFFICIENCY, ONE ROOM, ONE BATHROOM	EA PLUS SF	16,965.00 26.25		
	ONE BEDROOM, KITCHEN AND LIVING ROOM, ONE BATHROOM	EA PLUS SF	16,965.00 30.20		
	TWO BEDROOMS, KITCHEN AND LIVING ROOM, ONE BATHROOM	EA PLUS SF	16,965.00 31.71		
ALLOWANCES	ELIMINATE FURRING AND FOIL INSULATION AND INSTALL 3-1/2" BLANKET INSULATION ON EXISTING EXTERIOR STUD WALLS **DEDUCT** SF = SQUARE FOOT WALLS	SF	.47		
	ELIMINATE FURRING, INSULATION AND DRYWALL ON EXISTING PLASTERED EXTERIOR WALLS -- PATCH EXISTING PLASTER WHERE REQUIRED **DEDUCT** SF = SQUARE FOOT WALLS	SF	1.92		

CONVERSION — FULL HOUSE TO APARTMENTS

SPECIFICATIONS	
	CONVERT EXISTING ONE-FAMILY HOUSE TO FLATS (BASEMENT UNFINISHED) ACCORDING TO THE FOLLOWING SPECIFICATIONS:
PLANS & PERMIT	PREPARE PLANS AND OBTAIN PERMITS AS REQUIRED, PERMIT FEE COST **NOT** INCLUDED
TEAR-OUT AND DEMOLITION	GUT INTERIOR, LEAVING ONLY STUDS, FLOORS, STAIRS AND CHIMNEY REMOVE ALL PLUMBING, HEATING SYSTEM, WIRING AND ALL PLASTER EXCEPT ON INSIDE OF EXTERIOR MASONRY WALLS
MASONRY	INSTALL NEW FLUE IN EXISTING CHIMNEY: TEAR OUT PORTION OF CHIMNEY FROM INSIDE HOUSE, INSTALL NEW TERRA COTTA FLUE LINING, RE-BRICK
FRAMING	RE-BUILD STUD PARTITIONS ACCORDING TO PLANS
ROOFING, GUTTERS	INSTALL NEW ALUMINUM GUTTERS AND DOWNSPOUTS
DOORS	NEW EXTERIOR DOOR, 3-0 X 6-8, SOLID CORE, FLUSH SOLID CORE 1-3/4" DOORS AT ENTRANCE AND REAR OF EACH APARTMENT NEW INTERIOR DOORS, BIRCH FLUSH, 1-3/8" SOLID CORE 1-3/4" DOOR ON FURNACE ROOM ENCLOSURE TRIM ALL DOORS WITH RANCH CASING AND STOPS
WINDOWS	RE-TRIM ALL WINDOWS WITH RANCH CASING, STOPS, STOOL AND APRON
PLUMBING	IN EACH APARTMENT: BATHROOM WITH LAVATORY, TUB AND W.C., AND KITCHEN PLUMBING WITH NEW WATER PIPING AND STACK
HEATING & AIR CONDI-TIONING	NEW FORCED AIR HEAT AND COOLING SYSTEM, ELECTRIC OR GAS, INCLUDING HUMIDIFIER, ALL DUCTWORK, REGISTERS FOR EACH APARTMENT (IF HEATING BY GAS, VENT TO OUTSIDE)
ELECTRICAL	SEPARATE 100-AMP SERVICE FOR EACH APARTMENT WITH CIRCUIT BREAKERS RE-WIRE HOUSE TO CODE, INCLUDING WALL OUTLETS, SWITCHES, FIXTURES
INSULATION	6" BLANKET INSULATION IN TOP FLOOR CEILING 3-1/2" BLANKET INSULATION BETWEEN FLOORS ALUMINUM FOIL INSULATION ON FURRED EXTERIOR WALL
INTERIOR WALLS AND CEILINGS	5/8" FIRECODE DRYWALL ON ENTIRE BASEMENT CEILING, WALLS OF FURNACE ENCLOSURES, BASEMENT BEAMS AND COLUMNS, HALLWAYS 1/2" DRYWALL ON ALL OTHER WALLS AND CEILINGS, ALL DRYWALL FINISHED, 3 COATS, READY FOR PAINTING CERAMIC TILE ON WALLS OF BATHS, 4 FEET ABOVE THE FLOOR (6 FEET ABOVE IN BATHTUB AREAS)
MILLWORK, TRIM, STAIRS	RANCH BASE AND OAK SHOE THROUGHOUT HOUSE LINEN CLOSET 24" DEEP, 36" WIDE IN EACH APARTMENT CLOSET TRIM IN ALL BEDROOM CLOSETS REPAIR MAIN STAIRS, REPLACE HANDRAILS, BALUSTERS, ETC. NEW STAIRS TO BASEMENT, OR ELIMINATE AND CLOSE IN OPENING

CONVERSION — FULL HOUSE TO APARTMENTS

SPECIFICATIONS	
	SPECIFICATIONS FOR FULL HOUSE CONVERSION, CONTINUED:
CABINETS & APPLIANCES	INSTALL IN EACH APARTMENT: 12 LF OF UPPER AND LOWER KITCHEN CABINETS (RANGE, SINK AND REFRIGERATOR AREAS INCLUDED IN MEASUREMENT) BROOM CABINET COUNTERTOP, STAINLESS STEEL SINK, DISPOSAL NEW GAS OR ELECTRIC RANGE NEW GAS OR ELECTRIC HOT WATER HEATER DUCT TYPE KITCHEN FAN AND HOOD, VENTED TO OUTSIDE
SPECIALTIES	HINGED DOOR MEDICINE CABINETS IN BATHROOM WITH FLUORESCENT LIGHTS EACH SIDE OF MIRROR
FLOOR COVERING	CERAMIC TILE FLOOR IN BATHROOMS INLAID ROLL FLOORING IN KITCHENS $12/YD CARPET WITH PAD IN OTHER ROOMS
PAINTING	PAINT EXTERIOR TRIM, CORNICE, WINDOWS, DOORS, TWO COATS PAINT INTERIOR WALLS AND CEILINGS, DOORS, TRIM, TWO COATS
CLEAN-UP AND HAULING	CHUTE DEMOLITION RUBBISH TO TRUCK FROM BUILDING CLEAN UP TRASH DURING AND AT COMPLETION OF JOB DUMPSTER AND DUMPING FEE **NOT** INCLUDED

SPECIFICATIONS		UNIT	PRICE	LOCAL AREA MODIFICATION	
				%	PRICE
FULL HOUSE CONVERSION	TWO STORY HOUSE (2 APARTMENTS)	EA PLUS SF	30,720.00 44.10		
	THREE STORY HOUSE (3 APARTMENTS)	EA PLUS SF	46,050.00 44.10		
	FOUR STORY HOUSE (4 APARTMENTS) SF = TOTAL LIVING AREA OF HOUSE	EA PLUS SF	61,500.00 44.10		
ALLOWANCES	ELIMINATE FURRING AND FOIL INSULATION AND INSTALL 4" BLANKET INSULATION ON EXISTING EXTERIOR STUD WALLS **DEDUCT** SF = WALLS	SF	.47		
	ELIMINATE FURRING, INSULATION AND DRYWALL ON EXISTING PLASTERED EXTERIOR WALLS — PATCH EXISTING PLASTER WHERE REQUIRED **DEDUCT** SF = WALLS	SF	1.92		

RENOVATION — FULL HOUSE

SPECIFICATIONS	
	RENOVATE COMPLETE HOUSE ACCORDING TO THE FOLLOWING SPECIFICATIONS:
PLANS & PERMIT	PREPARE PLANS AND OBTAIN PERMITS AS REQUIRED, PERMIT FEE COST **NOT** INCLUDED
TEAR-OUT AND DEMOLITION	GUT INTERIOR, LEAVING ONLY STUDS, FLOORS, STAIRS AND CHIMNEY REMOVE ALL PLUMBING, HEATING SYSTEM, WIRING AND ALL PLASTER EXCEPT ON INSIDE OF EXTERIOR MASONRY WALLS
MASONRY	INSTALL NEW FLUE IN EXISTING CHIMNEY: TEAR OUT PORTION OF CHIMNEY FROM INSIDE HOUSE, INSTALL NEW TERRA COTTA FLUE LINING, RE-BRICK
FRAMING	RE-BUILD STUD PARTITIONS ACCORDING TO PLANS FURR EXTERIOR WALLS
ROOFING, GUTTERS	INSTALL NEW ALUMINUM GUTTERS AND DOWNSPOUTS
DOORS	NEW EXTERIOR DOOR, 3-0 X 6-8, SOLID CORE, FLUSH, 1-3/4" NEW REAR DOOR, 2-8 X 6-8, TWO PANELS, FOUR LIGHTS, 1-3/4" NEW PRE-HUNG INTERIOR DOORS, FLUSH, 1-3/8" TRIM ALL DOORS WITH RANCH CASING AND STOP, INCLUDING ALL HARDWARE
WINDOWS	RE-TRIM ALL WINDOWS WITH RANCH CASING, STOPS, STOOL AND APRON
PLUMBING	POWDER ROOM WITH LAVATORY AND W.C. TWO BATHS WITH LAVATORY, TUB AND W.C. NEW KITCHEN SINK AND DISPOSAL ALL NEW PLUMBING PIPES THROUGHOUT HOUSE NEW GAS OR ELECTRIC HOT WATER HEATER
HEATING & AIR CONDITIONING	NEW FORCED AIR HEAT AND COOLING SYSTEM, ELECTRIC OR GAS, INCLUDING HUMIDIFIER, ALL DUCTWORK, REGISTERS
ELECTRICAL	HEAVY UP SERVICE TO 200 AMPS WITH CIRCUIT BREAKERS RE-WIRE HOUSE TO CODE, INCLUDING OUTLETS, SWITCHES, FIXTURES
INSULATION	6" BLANKET INSULATION IN TOP FLOOR CEILING ALUMINUM FOIL INSULATION ON FURRED EXTERIOR WALL
INTERIOR WALLS AND CEILINGS	1/2" DRYWALL ON ALL WALLS AND CEILINGS, FINISHED 3 COATS, READY FOR PAINTING CERAMIC TILE ON WALLS OF TWO BATHS, 4 FEET ABOVE THE FLOOR AND 6 FEET ABOVE THE FLOOR IN BATHTUB AREA
MILLWORK, TRIM, STAIRS	RANCH BASE AND OAK SHOE THROUGHOUT HOUSE LINEN CLOSET 24" DEEP, 36" WIDE WITH SHELVES FULL DEPTH OF CLOSET CLOSET TRIM IN ALL BEDROOM CLOSETS REPAIR MAIN STAIRS, REPLACE HANDRAILS AND BALUSTERS

SPECIFICATIONS	
	SPECIFICATIONS FOR FULL HOUSE RENOVATION, CONTINUED:
CABINETS & APPLIANCES	15 LF OF UPPER AND LOWER KITCHEN CABINETS (RANGE, SINK AND REFRIGERATOR AREAS INCLUDED IN MEASUREMENT) BROOM CABINET COUNTERTOP, STAINLESS STEEL SINK, DISPOSAL NEW GAS OR ELECTRIC RANGE REFRIGERATOR DUCT TYPE KITCHEN FAN AND HOOD, VENTED TO OUTSIDE CLOTHES WASHER/DRYER
SPECIALTIES	HINGED DOOR MEDICINE CABINETS IN BATHROOMS AND POWDER ROOM WITH FLUORESCENT LIGHTS EACH SIDE OF MIRRORS
FLOOR COVERING	CERAMIC TILE FLOOR IN BATHROOMS VINYL COMPOSITION TILE IN POWDER ROOM AND KITCHEN $12/YD CARPET WITH PAD IN OTHER ROOMS
PAINTING	PAINT EXTERIOR TRIM, CORNICE, WINDOWS, DOORS, TWO COATS PAINT INTERIOR WALLS AND CEILINGS, DOORS, TRIM, TWO COATS
CLEAN-UP AND HAULING	CHUTE DEMOLITION RUBBISH TO TRUCK FROM BUILDING CLEAN UP TRASH DURING AND AT COMPLETION OF JOB DUMPSTER AND DUMPING FEE **NOT** INCLUDED

SPECIFICATIONS		UNIT	PRICE	LOCAL AREA MODIFICATION	
				%	PRICE
FULL HOUSE RENOVATION	ONE STORY HOUSE, 2 BATHS	EA PLUS SF	49,050.00 24.63		
	TWO STORY HOUSE, POWDER ROOM AND 2 BATHS	EA PLUS SF	39,546.00 31.95		
	THREE STORY HOUSE, POWDER ROOM & 4 BATHS SF = TOTAL LIVING AREA OF HOUSE (**NOT** INCLUDING BASEMENT UNLESS IT INCLUDES FINISHED LIVING AREA)	EA PLUS SF	43,155.00 31.95		
ALLOWANCES	ELIMINATE FURRING AND FOIL INSULATION AND INSTALL 3-1/2" BLANKET INSULATION ON EXISTING EXTERIOR STUD WALLS **DEDUCT** SF = WALLS	SF	.47		
	ELIMINATE FURRING, INSULATION AND DRYWALL ON EXISTING PLASTERED EXTERIOR WALLS — PATCH EXISTING PLASTER WHERE REQUIRED **DEDUCT** SF = WALLS	SF	1.92		

PORCH AND SLAB ALTERATIONS

SPECIFICATIONS		UNIT	PRICE	LOCAL AREA MODIFICATION	
				%	PRICE
REMOVE ROOF AND FLOOR	• REMOVE EXISTING WOOD PORCH, INCLUDING ALL ROOF AND FLOOR FRAMING AND STEPS, REMOVE DEBRIS FROM PREMISES • PATCH BRICKWORK ON FRONT WHERE WOOD BEAMS ATTACH TO HOUSE, TOOTHING IN TO MATCH EXISTING BRICKWORK AS CLOSELY AS POSSIBLE • DIG, FORM & POUR REINFORCED CONCRETE PLATFORM WITH 4-0 WIDE STEPS SF = TOP SURFACE NEW PLATFORM & STEPS	EA PLUS SF	1200.00 25.50		
REMOVE ROOF	REMOVE EXISTING PORCH ROOF, INCLUDING SUPPORT POSTS AND SCREENING — PATCH BRICKWORK ON FRONT WHERE ROOF ATTACHES TO HOUSE, TOOTHING IN TO MATCH EXISTING AS CLOSELY AS POSSIBLE SF = PORCH ROOF	EA PLUS SF	300.00 3.00		
REPLACE RAILING	REMOVE EXISTING WOOD RAILING, INSTALL ECONOMY GRADE ORNAMENTAL IRON RAILING LF = RAILING	LF	30.00		
REMOVE FLOOR AND REPLACE	REMOVE EXISTING PORCH FLOOR, LEAVING PILLARS TO SUPPORT ROOF, REMOVE DEBRIS FROM PREMISES — DIG, FORM AND POUR REINFORCED CONCRETE PLATFORM WITH 4-0 WIDE STEPS SF = TOP SURFACE NEW PLATFORM AND STEPS	EA PLUS SF	1245.00 20.40		
REPAIR METAL PLATFORM	WELD BROKEN PARTS TO METAL PORCH, LEVEL PLATFORM, STEPS AND RAILING AND ANCHOR IN PLACE	EA	231.00		
MUDJACK	RAISE EXISTING SETTLED CONCRETE WALKWAY OR SLAB BY FILLING VOID UNDER WITH PUMPED CONCRETE GROUT: • DRILL HOLES IN CONCRETE SLAB • PUMP CONCRETE GROUT UNDER PRESSURE THRU HOLES, FILLING VOID UNDER SLAB AND COMPACTING EARTH BELOW • WHEN SLAB RAISED AS MUCH AS 8" TO ORIGINAL LEVEL, PATCH DRILLED HOLES WITH MORTAR MIX SF = SLAB	EA PLUS SF	885.00 2.63		

SPECIFICATIONS		UNIT	PRICE	LOCAL AREA MODIFICATION	
				%	PRICE
AREAWAY FOR BASEMENT APARTMENT	• DIG OUT AND REMOVE DIRT AS REQUIRED FOR AREAWAY 5 FEET DEEP AND UP TO 18 FEET WIDE • POUR CONCRETE FOOTINGS FOR 8" MASONRY WALL • FORM AND POUR REINFORCED (6 X 6 WWM) CONCRETE SLAB 4" THICK WITH BASEMENT DRAIN CONNECTED WITH SEWER • INSTALL CONCRETE STEPS TO GRADE, UP TO 8 STEPS 4 FEET WIDE • INSTALL 4" BRICK AND 4" BLOCK WALL WITH TOP ROLOK COURSE AROUND PERIMETER OF AREAWAY WALL EXTENDING FROM FOOTINGS TO GRADE SF = AREAWAY FLOOR	EA PLUS SF	900.00 57.00		
CLEAR AREAWAY	REMOVE OBSTRUCTION FROM BASEMENT AREAWAY DRAIN	EA	150.00		
WEEP HOLES IN EXISTING RETAINING WALL	• CUT 1 SF HOLE IN EXISTING RETAINING WALL • DIG OUT TO HOLE BEHIND WALL AND INSTALL 1 CF CRUSHED STONE • GROUT IN 3" PLASTIC PIPE				
	8" BRICK & BLOCK	EA	39.00		
	12" BRICK & BLOCK	EA	45.00		
	12" SOLID BLOCK	EA	51.00		
	EA = EACH WEEPHOLE				
FRENCH DRAIN	• BREAK CONCRETE BASEMENT FLOOR ALONG INSIDE OF FOOTINGS AND EXCAVATE TO DEPTH OF FOOTINGS • INSTALL PLASTIC PIPE ALONG INSIDE OF FOOTINGS WITH 1/8" SLOPE TO SUMP IN BASEMENT FLOOR • DISCHARGE WATER WITH 1/3 HP SUMP PUMP TO STORM SEWER OR OUTSIDE TO DAYLIGHT • BACKFILL AND PATCH FLOOR LF = FOOTINGS	EA PLUS LF	789.00 30.00		

FIREWALL, SOUND BARRIER, SHOWER PAN

SPECIFICATIONS		UNIT	PRICE	LOCAL AREA MODIFICATION	
				%	PRICE
FIREWALL AROUND BOILER & HOT WATER HEATER	BUILD WALL ENCLOSING 6' X 8' SPACE AT END OF EXISTING BASEMENT • 8" X 8" X 16" BLOCK **OR** 2 X 4 STUD WALL AND 5/8" FIRECODE DRYWALL BOTH SIDES OF WALL • CLASS B FIREDOOR, APPROX. 2-6 X 6-8, WITH AUTOMATIC CLOSER • 5/8" FIRECODE DRYWALL ON CEILING, TAPED AND SPACKLED, FILL IN AROUND PIPES AS THOROUGHLY AS POSSIBLE • 8" X 12" DUCT FROM FURNACE ROOF 10 FEET OR LESS TO OUTSIDE FOR VENTILATION • PULL CHAIN CEILING LIGHT	EA	2700.00		
SOUND BARRIER	INSTALL SOUND BARRIER BETWEEN BASEMENT CEILING AND FIRST FLOOR APARTMENT WITH FULL THICK BLANKET INSULATION, RESILIENT CHANNEL 24" O.C. AND 1/2" DRYWALL, TAPED AND SPACKLED, READY FOR PAINTING	SF	6.00		
REPLACE LEAD SHOWER PAN	REPLACE EXISTING LEAD PAN IN SHOWER STALL • TEAR OUT TILE FLOOR AND WALLS UP THREE ROWS FROM BASE • SUPPLY AND INSTALL NEW DRAIN ASSEMBLY AND COVER • REPLACE EXISTING LEAD PAN WITH VINYL OR RUBBER PAN • INSTALL NEW TILE ON FLOOR AND UP THREE ROWS IN MUD, TILE TO MATCH EXISTING AS CLOSELY AS POSSIBLE	EA	1044.00		

SPECIFICATIONS		UNIT	PRICE	LOCAL AREA MODIFICATION %	LOCAL AREA MODIFICATION PRICE
CLOSE IN EXTERIOR DOORWAY OR WINDOW	REMOVE DOOR OR WINDOW AND CLOSE IN WITH STUDS, SHEATHING, SIDING, INSULATION, FURRING AND/OR MASONRY AS REQUIRED — FINISH INSIDE WITH DRYWALL AND BASE AND SHOE IF REQUIRED — TOOTH IN BRICK TO EXISTING WHEN BRICK IS SPECIFIED				
	FRAMED WALL WITH SIDING	EA	462.00		
	FRAME WITH BRICK VENEER	EA	909.00		
	BRICK AND BLOCK	EA	1080.00		
CLOSE IN INTERIOR DOORWAY	REMOVE EXISTING INTERIOR DOORWAY AND CLOSE IN WALL WITH STUDS AND DRYWALL AND BASE AND SHOE BOTH SIDES OF WALL	EA	420.00		
CLOSE IN FIREPLACE	CLOSE IN EXISTING FIREPLACE WITH STUDS, DRYWALL AND BASE	EA	180.00		
	SAME AS ABOVE, AND REMOVE BRICK FACING	EA	261.00		
REMOVE FRONT HEARTH	REMOVE FRONT HEARTH AND PATCH WITH HARDWOOD FLOORING, SANDED AND FINISHED	EA	171.00		
CLOSE IN TRANSOM	REMOVE EXISTING TRANSOM OVER DOOR AND CLOSE IN WITH DRYWALL	EA	81.00		
PLASTER PATCHING	IF ANY OF THE ABOVE CLOSE-INS REQUIRE PLASTER PATCHING **ADD**	EA	111.00		

BREAK THROUGH WALLS FOR NEW DOORWAYS

SPECIFICATIONS		UNIT	PRICE	LOCAL AREA MODIFICATION	
				%	PRICE
NEW DOOR IN EXTERIOR WALL	BREAK OPENING IN EXISTING FRAME WALL, INSTALL HEADER OR LINTEL AS REQUIRED AND INSTALL DOOR, INCLUDING FRAME, TRIM AND HARDWARE				
	SOLID CORE BIRCH, PRE-HUNG, 3-0 X 6-8	EA	1194.00		
	COLONIAL SIX-PANEL, PRE-HUNG, 3-0 X 6-8	EA	1383.00		
	4-LIGHT SIDE OR REAR DOOR, 2-8 X 6-8	EA	1329.00		
	WOOD GLIDER, BUILDING QUALITY, 6-0 X 6-8	EA	2169.00		
	ALUMINUM GLIDER, 6-0 X 6-8	EA	1710.00		
	SAME AS ABOVE, REMOVE EXISTING WINDOW AND WALL UNDER, OMIT HEADER OR LINTEL AND INSTALL DOOR **DEDUCT**	EA	144.00		
	SAME AS ABOVE, IN BRICK VENEER WALL **ADD**	EA	57.00		
	SAME AS ABOVE, IN BRICK AND BLOCK WALL **ADD**	EA	90.00		
NEW DOOR IN INTERIOR WALL	BREAK THROUGH EXISTING NON-BEARING INTERIOR WALL AND HANG DOOR IN OPENING, INCLUDING DOOR, STOPS, TRIM, HINGES AND PASSAGE SET				
	BIRCH FLUSH, HOLLOW CORE, 2-8 X 6-8	EA	606.00		
	6 PANEL COLONIAL, 2-8 X 6-8	EA	732.00		
DOUBLE DOORS	DOUBLE DOORS: DOUBLE COST OF ONE DOOR AND **ADD**	EA	75.00		
	SAME AS ABOVE IN LOAD-BEARING WALL **ADD**	EA	210.00		
REMOVE WALL AND MAKE CASED OPENING	BREAK THROUGH EXISTING BEARING OR NON-BEARING WALL AND MAKE NEW CASED OPENING WITH JAMBS AND 2 SIDES OF DOOR TRIM				
	Non-Bearing Wall				
	3-0 OPENING	EA	516.00		
	6-0 OPENING	EA	654.00		
	Bearing Wall				
	3-0 OPENING	EA	708.00		
	6-0 OPENING	EA	990.00		

SPECIFICATIONS		UNIT	PRICE	LOCAL AREA MODIFICATION %	LOCAL AREA MODIFICATION PRICE
REMOVE WALL AND MAKE OPENING	BREAK THROUGH EXISTING LOAD-BEARING INTERIOR WALL AND MAKE NEW OPENING WITH MINIMUM OF 6" WALL SHOWING ON EACH SIDE AND 6" HEADER SHOWING ABOVE, PATCH WALLS, CEILING, MOULDINGS & FLOOR TO MATCH EXISTING, 10 TO 16 LF OPENING EA = EACH JOB	EA	1080.00		
	SAME AS ABOVE, WITH FLUSH CEILING AND 6" MINIMUM WALLS SHOWING EACH SIDE **ADD**	EA	225.00		
	SAME AS ABOVE, WITH FLUSH CEILING AND FLUSH WALLS **ADD**	EA	399.00		
DOOR ADJUSTMENT	CUT OFF BOTTOM OF SOLID WOOD DOOR FOR CARPETING CLEARANCE				
	INTERIOR	EA	18.00		
	EXTERIOR	EA	63.00		
	CUT OFF BOTTOM RAIL OF HOLLOW CORE DOOR AND RAISE BOTTOM RAIL INSIDE THE VENEER	EA	33.00		
	RE-SET SCREWS IN HINGES EA = UP TO 3 HINGES ON ONE DOOR	EA	18.00		
	SAND OR PLANE DOOR TO FIT EXISTING OPENING	EA	18.00		
	ADD HINGE IN MIDDLE OF DOOR TO CORRECT WARP	EA	18.00		
	ALTER SWING OF DOOR, RE-USING SAME HARDWARE	EA	45.00		
	REMOVE EXISTING LOCKSET AND REPLACE WITH NEW LOCKSET WITHOUT BORING NEW HOLES, PASSAGE OR PRIVACY @ $6	EA	27.00		
ADJUST POCKET DOOR	OPEN WALL, UNDERCUT, RESET, ADJUST AND LUBRICATE POCKET DOOR, CLOSE WALL WITH DRYWALL AND INSTALL NEW ONE-PIECE BASE AND CASING	EA	147.00		
	FOR PLASTERING WALL ABOVE INSTEAD OF DRYWALL **ADD**	EA	9.75		
	SUBSTITUTE 1" X 4" BASE, OG AND SHOE IN ABOVE CLOSE-IN **ADD**	EA	9.00		

NEW WINDOWS IN EXISTING WALLS

SPECIFICATIONS		UNIT	PRICE	LOCAL AREA MODIFICATION	
				%	PRICE
REPLACE EXISTING DOOR WITH NEW WINDOW	• REMOVE EXISTING DOOR AND FRAME FROM FRAME WALL • INSTALL NEW BUILDER QUALITY WINDOW AND FRAME NOT MORE THAN WIDTH OF OLD DOOR • CLOSE IN BELOW NEW WINDOW WITH STUDS, SHEATHING AND BEVELED WOOD OR ALUMINUM SIDING • 1/2" DRYWALL, PLASTER OR PLYWOOD PANELING ON INSIDE OF WALL • INSTALL BASE AND SHOE ON INSIDE OF WALL TO MATCH EXISTING AS CLOSELY AS POSSIBLE				
	WOOD DOUBLE HUNG, 3-0 X 4-6	EA	783.00		
	WOOD CASEMENT, 2-4 X 5-6	EA	750.00		
	WOOD SLIDER, 3-0 X 3-6	EA	732.00		
	ALUMINUM SLIDER, 3-0 X 4-0	EA	624.00		
	SAME AS ABOVE WITH DIFFERENT EXTERIOR WALLS				
	BRICK VENEER **ADD**	EA	114.00		
	BRICK & BLOCK **ADD**	EA	129.00		
NEW WINDOW IN EXISTING WALL	BREAK OPENING IN EXISTING EXTERIOR FRAME WALL, INSTALL HEADER, PATCH OPENING INCLUDING SIDING AND INTERIOR WALL COVERING AND INSTALL NEW WINDOW				
	WOOD DOUBLE HUNG, 3-0 X 4-6	EA	855.00		
	WOOD CASEMENT, 2-4 X 5-6	EA	837.00		
	WOOD SLIDER, 3-0 X 3-6	EA	693.00		
	ALUMINUM SLIDER, 3-0 X 4-0	EA	651.00		
	SAME AS ABOVE WITH DIFFERENT EXTERIOR WALLS				
	BRICK VENEER **ADD**	EA	15.00		
	BRICK & BLOCK **ADD**	EA	45.00		

WINDOW ALTERATIONS & ADJUSTMENTS

SPECIFICATIONS		UNIT	PRICE	LOCAL AREA MODIFICATION %	LOCAL AREA MODIFICATION PRICE
RE-SET WINDOW	REMOVE EXISTING WOOD WINDOW, RE-SET AND INSTALL NEW INTERIOR TRIM	EA	84.00		
REPLACE GLASS	REMOVE OLD BROKEN GLASS AND PUTTY AND RE-GLAZE WOOD WINDOW PANE				
	UP TO 8" X 12"	EA	24.00		
	UP TO 36" X 24"	EA	39.00		
	REMOVE OLD BROKEN GLASS AND PUTTY FROM METAL WINDOW AND RE-GLAZE				
	WITH STANDARD GLASS	EA	36.00		
	WITH PLATE OR WIRED	EA	51.00		
	REPLACE BROKEN STORM WINDOW PANE	EA	36.00		
STUCK WINDOW	FREE STUCK WINDOW TO OPERATE FREELY AND MOVE STOPS AS NECESSARY	EA	27.00		
SASH REPAIRS	INSTALL SASH FASTENER ON WINDOW	EA	7.50		
	KEYLOCK TYPE FASTENER ON EXISTING SASH	EA	27.00		
	REPLACE CASEMENT WINDOW CRANK	EA	15.00		
	REPLACE ROTTED EXTERIOR WINDOW SILL	EA	48.00		
	REPLACE EXISTING SASH CORDS WITH SASH CHAINS EA = EACH SASH CORD	EA	30.00		
	REPLACE WINDOW STOPS AND PARTING BEADS ON EXISTING WINDOW EA = EACH WINDOW	EA	36.00		

SKYLIGHT IN EXISTING ROOF

SPECIFICATIONS		UNIT	PRICE	LOCAL AREA MODIFICATION	
				%	PRICE
FIXED SKYLIGHT	• OPEN EXISTING ASPHALT OR FIBERGLASS SHINGLED ROOF • INSTALL HEADERS ON EXISTING RAFTERS • WOOD CURB WITH 2" X 6" • ALUMINUM FLASHING • PATCH ROOFING AS REQUIRED • FINISH INTERIOR WITH WOOD TRIM AND/OR DRYWALL TAPED AND FINISHED (**NOT** PAINTED) • WOOD FRAMED SKYLIGHT WITH DOUBLE GLAZED TEMPERED GLASS, EXTERIOR OF FRAME ALUMINUM CLAD WITH BAKED ON ENAMEL FINISH • CORD OPERATED ROLLER BLIND ***Outside Frame***				
	30" X 22"	EA	894.00		
	X 30"	EA	954.00		
	45-1/2" X 30"	EA	1074.00		
	X 45-1/2"	EA	1230.00		
VENTILATING SKYLIGHT	SAME AS ABOVE, VENTILATING SKYLIGHT WITH INSECT SCREEN **ADD**	EA	69.00		
SKYLIGHT ACCESSORIES	ROD CONTROL FOR BLINDS AND/OR WINDOWS	EA	63.00		
	ELECTRIC WINDOW MOTOR CONTROL	EA	330.00		
LIGHT SHAFT	BUILD LIGHT SHAFT FROM SLOPED RAFTERS TO FLAT CEILING BELOW • INSTALL HEADERS ON EXISTING CEILING JOISTS • FRAME OUT LIGHT SHAFT FROM EXISTING RAFTER HEADERS TO NEW JOIST HEADERS • INSULATE LIGHT SHAFT • FINISH INTERIOR OF LIGHT SHAFT WITH DRYWALL, TAPED AND FINISHED (**NOT** PAINTED) TO FIXED AND VENTILATING SKYLIGHT PRICES **ADD** AS SHOWN BELOW: ***Distance from Room's Finished Ceiling To Closest Point of Roof Above***				
	UP TO 2'-0" **ADD**	EA	354.00		
	2'-0" TO 4'-0" **ADD**	EA	480.00		
	4'-0" TO 6'-0" **ADD**	EA	600.00		
	6'-0" TO 8'-0" **ADD**	EA	735.00		
	8'-0" TO 10'-0" **ADD**	EA	855.00		

SPECIFICATIONS			UNIT	PRICE	LOCAL AREA MODIFICATION %	LOCAL AREA MODIFICATION PRICE
ROOF WINDOW IN EXISTING ROOF	• OPEN EXISTING ASPHALT OR FIBERGLASS SHINGLED ROOF • INSTALL HEADERS ON EXISTING RAFTERS • INSTALL WOOD CURB WITH 2" X 6" • ALUMINUM FLASHING • PATCH ROOFING AS REQUIRED • FINISH INTERIOR WITH WOOD TRIM AND/OR DRYWALL, TAPED AND FINISHED (**NOT** PAINTED) • WOOD FRAMED ROOF WINDOW WITH DOUBLE GLAZED TEMPERED GLASS, EXTERIOR OF FRAME ALUMINUM CLAD WITH BAKED ON ENAMEL FINISH, INTERIOR OF FRAME UNFINISHED WOOD • INSECT SCREEN • CENTER PIVOT					
	Outside Frame					
	21-1/2" X 27-1/2"		EA	984.00		
	X 38-1/2"		EA	1035.00		
	27-1/2" X 46-1/2"		EA	1185.00		
	30-5/8" X 38-1/2"		EA	1140.00		
	X 55"		EA	1284.00		
	36-7/8" X 62-7/8"		EA	1470.00		
	44-3/4" X 46-1/2"		EA	1434.00		
	52-5/8" X 38-1/2"		EA	1470.00		
	X 55"		EA	1515.00		
	SAME AS ABOVE, TOP HUNG WITH SILL MOUNTED SCISSOR TYPE OPERATOR					
	Outside Frame					
	21-1/2" X 38-1/2"		EA	1059.00		
	30-5/8" X 38-1/2"		EA	1194.00		
	X 55"		EA	1329.00		
	44-3/4" X 46-1/2"		EA	1485.00		
INSTALL ROOF WINDOW OR SKYLIGHT IN ROOF WITH OTHER ROOF COVERING	BUILT-UP	**ADD**	EA	384.00		
	SLATE	**ADD**	EA	339.00		
	CEDAR SHAKES AND SHINGLES	**ADD**	EA	153.00		
	CLAY TILE	**ADD**	EA	399.00		
	CONCRETE ROOFING TILES	**ADD**	EA	315.00		

FRAMING ALTERATIONS

SPECIFICATIONS		UNIT	PRICE	LOCAL AREA MODIFICATION	
				%	PRICE
REPLACE COLUMN	REPLACE EXISTING WOOD OR MASONRY COLUMN WITH NEW 3" HOLLOW STEEL COLUMN WITH BASE AND CAP, UP TO 8'-0" LONG — INCLUDES TEAR-OUT OF OLD COLUMN AND SHORING	EA	177.00		
REPLACE WOOD BEAM	REPLACE EXISTING WOOD BEAM WITH NEW 8" #13 STEEL BEAM — INCLUDES TEAR-OUT OF OLD BEAM	LF	603.00		
REPLACE SILL PLATE	REMOVE DEFECTIVE SILL PLATE AND REPLACE WITH NEW 2" X 6" OR 4" X 6" SILL — INCLUDES JACKING UP WALL, BUT **NO** WALL PATCHING ABOVE				
	IN BASEMENT WITH GOOD HEADROOM AND EASY ACCESS	LF	65.70		
	IN CRAWL SPACE WITH AT LEAST 18" HEAD-ROOM AND EASY ACCESS	LF	103.20		
DOUBLE UP JOISTS (SISTERING)	DOUBLE UP EXISTING JOISTS WITH NEW JOISTS SPIKED OR BOLTED TO EXISTING JOISTS, EASY ACCESS				
	2" X 6"	LF	3.57		
	2" X 8"	LF	4.10		
	LF = NEW JOISTS				
DOUBLE UP RAFTERS (SISTERING)	DOUBLE UP EXISTING RAFTERS WITH NEW 2 X 6 RAFTERS SPIKED OR BOLTED TO EXISTING RAFTERS				
	GROUND FLOOR RAFTERS	LF	3.12		
	SECOND FLOOR RAFTERS	LF	3.57		
	THIRD FLOOR RAFTERS	LF	4.02		
DOUBLE UP RIDGE-BOARD	DOUBLE UP RIDGEBOARD WITH NEW RIDGE-BOARD SPIKED OR BOLTED TO EXISTING RIDGEBOARD				
	2" X 10"	LF	3.08		
	2" X 12"	LF	3.90		
STRUCTURAL RIDGE BEAM	ADD TO EXISTING RIDGEBOARD, SPIKED OR BOLT-ED				
	(2) 2" X 10"	LF	5.85		
	(2) 2" X 12"	LF	7.49		
	STRUCTURAL RIDGE BEAM INSTALLED BELOW EXISTING RIDGEBOARD WITH 2" X 4" BRACES FROM ALL RAFTERS TO RIDGE BEAM				
	(2) 2" X 10"	LF	8.18		
	(3) 2" X 10"	LF	10.83		
	(2) 2" X 12"	LF	9.68		
	(3) 2" X 12"	LF	13.19		

HomeTech
Remodeling and Renovation
Cost Estimator

Section II

1. PLANS AND PERMITS

1

SPECIFICATIONS		UNIT	PRICE	LOCAL AREA MODIFICATION %	LOCAL AREA MODIFICATION PRICE
ALL REMODELING AND RENOVATION WORK	IN ADDITION TO BUILDING PERMIT FEE. COST AND PRICE ARE BASED ON TOTAL AMOUNT OF JOB FOUNDATION PLANS, FLOOR PLANS, ELEVATIONS, SECTIONS, AS REQUIRED INCLUDES ON-SITE SURVEY & MEASUREMENTS BY DESIGNER, BUT DOES **NOT** INCLUDE CONFERENCES WITH OR APPROVAL OF PLANS BY OWNER EA = TOTAL JOB PER $M = PER $1,000 OF TOTAL JOB COST (CONTRACTOR'S COST) OR JOB PRICE (PRICE TO CUSTOMER)	EA PLUS PER $M	300.00 10.00		
	SAME AS ABOVE AND INCLUDING CONFERENCES WITH AND APPROVAL OF PLANS BY OWNER	EA PLUS PER $M	900.00 55.00		
	SAME AS ABOVE PLUS SELECTION OF PRIME CONTRACTOR **ADD**	PER $M	6.00		
	SAME AS ABOVE PLUS ARCHITECTURAL SUPERVISION TO COMPLETION **ADD**	PER $M	25.00		
	FULL FEE FOR ALL OF ABOVE SERVICES	EA PLUS PER $M	900.00 96.00		
BUILDING PERMIT (AVERAGE AMOUNT FOR U.S.)	BUILDING PERMIT FEE, BASED ON TOTAL AMOUNT OF JOB PER $M = PER $1,000 OF TOTAL JOB COST (CONTRACTOR'S COST) OR JOB PRICE (PRICE TO CUSTOMER)	EA PLUS PER $M	75.00 10.00		

SPECIFICATIONS		UNIT	PRICE	LOCAL AREA MODIFICATION	
				%	PRICE
NOTE	ARCHITECTURAL SUPERVISION INCLUDES INSPECTION OF WORK AT LEAST EVERY OTHER DAY WHILE JOB IS PROGRESSING, APPROVAL OF PAYMENTS TO CONTRACTOR AND ON-THE-JOB CONFERENCES WITH CONTRACTOR AND CUSTOMER WHEN REQUIRED. NO DIRECT SUPERVISION OF WORKERS OR PREPARATION OF MATERIAL LISTS IS INCLUDED. A RECOMMENDED ARCHITECTURAL FEE FOR FULL SERVICE IS 13% OF TOTAL JOB PRICE (THE AMOUNT THAT THE CUSTOMER PAYS FOR THE WORK).				

2. PAVING AND SIDEWALK DEMOLITION

2

SPECIFICATIONS		UNIT	PRICE	LOCAL AREA MODIFICATION %	LOCAL AREA MODIFICATION PRICE
	NOTE: TEAR-OUT AND DEMOLITION PRICES INCLUDE PILING OF DEBRIS ON SITE BUT DO **NOT** INCLUDE REMOVAL FROM PREMISES. FOR REMOVAL OF DEBRIS FROM PREMISES, SEE "CLEAN-UP" ON PAGE 285.				
BITUMINOUS PAVING	REMOVE DRIVEWAY WITH SHOVEL OR LOADER	EA PLUS SF	463.50 .48		
CONCRETE PAVING	6" THICK CONCRETE DRIVEWAY REMOVED: WITH PNEUMATIC TOOL	SF	1.71		
	WITH SHOVEL OR BULLDOZER	EA PLUS SF	163.50 .48		
CURB AND GUTTER	WITH COMPRESSOR AND LABORERS, BREAK OUT 12" HIGH, 6" THICK CONCRETE CURB	EA PLUS SF	192.00 4.07		
SIDEWALK	BREAK UP BRICK OR CONCRETE SIDEWALK WITH PNEUMATIC TOOL	SF	1.68		
	BREAK UP FLAGSTONE SET IN SAND OR CRUSHED STONE, BY HAND	SF	1.58		
	BREAK UP FLAGSTONE SET IN MORTAR, WITH PNEUMATIC TOOL	SF	1.65		

SPECIFICATIONS		UNIT	PRICE	LOCAL AREA MODIFICATION %	LOCAL AREA MODIFICATION PRICE
CONCRETE FOOTING	REMOVE CONCRETE FOOTING WITH PNEUMATIC TOOL				
	8" X 16"	LF	3.87		
	12" X 24"	LF	7.29		
CONCRETE SLAB	BREAK UP EXISTING SLAB IN BASEMENT WITH PNEUMATIC TOOL AND HAUL RUBBLE OUTSIDE TO GRADE				
	2"	SF	2.01		
	4"	SF	2.34		
	6"	SF	2.75		
	BREAK UP EXISTING SLAB OUTSIDE ON GRADE WITH PNEUMATIC TOOL, 6" THICK				
	NO REINFORCEMENT	SF	1.83		
	REINFORCED	SF	2.61		
SUSPENDED SLAB AND STEPS	BREAK UP EXISTING SUSPENDED SLAB AND STEPS WITH PNEUMATIC TOOL, NOT OVER 10 FEET ABOVE GRADE, UP TO 4" THICK SF = TOTAL SLAB AND STEP AREA	SF	2.10		
CONCRETE WALL	REMOVE CONCRETE WALL WITH PNEUMATIC TOOL, 12" WALL				
	NO REINFORCEMENT	SF	6.17		
	REINFORCED	SF	9.17		
CONCRETE AND/OR BRICK PLATFORM & STEPS	BREAK UP EXISTING BRICK OR CONCRETE PLAT-FORM AND STEPS WITH PNEUMATIC TOOL (SOLID SIDES AND FILLED WITH RUBBLE)	SF	5.19		
CONCRETE SAWING	CUT CONCRETE WITH GAS CONCRETE SAW, PER INCH DEEP				
	CONCRETE SLAB WITH MESH	LF	2.40		
	CONCRETE WALL	LF	6.60		
	BRICK	LF	5.97		
	BLOCK	LF	4.95		

2. MASONRY DEMOLITION

2

SPECIFICATIONS		UNIT	PRICE	LOCAL AREA MODIFICATION %	LOCAL AREA MODIFICATION PRICE
BLOCK WALL	REMOVE BLOCK WALL BY HAND, **NO** REINFORCEMENT				
	4"	SF	.92		
	8"	SF	1.16		
	12"	SF	1.47		
BRICK WALL	REMOVE BRICK WALL BY HAND				
	4" VENEER	SF	1.70		
	8" SOLID	SF	4.68		
	12" SOLID	SF	7.34		
	REMOVE BRICK WALL WITH PNEUMATIC TOOL OR LARGE ELECTRIC HAMMER				
	4" VENEER	SF	1.70		
	8" SOLID	SF	3.92		
	12" SOLID	SF	5.73		
BRICK & BLOCK WALL	REMOVE BRICK AND BLOCK WALL WITH PNEUMATIC TOOL				
	8"	SF	2.69		
	12"	SF	3.81		
STONE WALL	REMOVE STONE WALL WITH PNEUMATIC TOOL, MORTAR JOINTS, 12" - 16" WALL	SF	7.88		

SPECIFICATIONS		UNIT	PRICE	LOCAL AREA MODIFICATION %	LOCAL AREA MODIFICATION PRICE
REMOVE MASONRY WALL AND INSTALL STEEL OVER	REMOVE ENTIRE BRICK OR BRICK AND BLOCK WALL (ONE STORY) AND INSTALL STEEL ANGLES OVER				
	2 STEEL ANGLES	LF	87.00		
	3 STEEL ANGLES	LF	96.00		
	SAME AS ABOVE AND INSTALL STEEL BEAM OR FLITCH PLATE SUPPORT	LF	96.00		
	IF STEEL COLUMNS ARE REQUIRED FOR STEEL BEAM OR FLITCH PLATE SUPPORT **ADD**	EA	114.00		
	EA = EACH COLUMN				
NEEDLE & SHORE	NEEDLE & SHORE MASONRY WALL OPENING, INCLUDING WALL BREAKTHROUGH AND REMOVAL				
	8" BLOCK	LF	124.50		
	4" BRICK & 8" BLOCK	LF	160.50		
	8" BRICK	LF	147.00		
	12" BRICK	LF	187.50		
	SAME AS ABOVE AND INSTALL 3 STEEL ANGLES **ADD**	LF	52.05		
	SAME AS ABOVE AND INSTALL STEEL BEAM **ADD**				
	8" WF 13#	LF	39.00		
	8" WF 17#	LF	51.00		
CHIMNEY	TEAR OUT EXISTING CHIMNEY FROM ROOF TO FOOTING, **NO** FLOOR OR WALL PATCHING				
	SMALL CHIMNEY, UP TO 20" X 20"				
	EXTERIOR	LF Down	11.25		
	INTERIOR	LF Down	14.48		
	LARGE CHIMNEY, 20" X 20" TO 48" X 48"				
	EXTERIOR	LF Down	17.70		
	INTERIOR	LF Down	27.00		
FIREPLACE	REMOVE BRICK FIREPLACE WITH 30" X 24" OPENING				
	SOFT MORTAR	EA	180.00		
	HARD MORTAR	EA	243.00		

2. EXTERIOR TEAR-OUT

2

SPECIFICATIONS		UNIT	PRICE	LOCAL AREA MODIFICATION	
				%	PRICE
REMOVE FRAME WALL AND INSTALL WOOD HEADER OR STEEL OVER	REMOVE ENTIRE WALL (ONE STORY) AND INSTALL WOOD HEADER OVER AS REQUIRED EXTERIOR WALL MATERIALS WOOD, COMPOSITION, STUCCO OR METAL SIDING OVER STUDS AND SHEATHING INTERIOR WALL MATERIALS DRYWALL, PLASTER AND LATH OR WOOD PANELING	LF	48.90		
	SAME AS ABOVE AND INSTALL STEEL BEAM OR FLITCH PLATE SUPPORT	LF	93.00		
EXTERIOR WALL COVERING	REMOVE SIDING FROM EXISTING EXTERIOR WALL				
	DROP OR BEVELED WOOD	SF	.51		
	PLYWOOD	SF	.38		
	BOARD AND BATTEN	SF	.47		
	CEDAR SHINGLES	SF	.47		
	ALUMINUM OR VINYL	SF	.47		
	STUCCO & WOOD LATH	SF	.56		
	STUCCO & METAL LATH	SF	.89		
EXTERIOR TRIM	FASCIA BOARD OR SOFFIT				
	1" X 6"	LF	.59		
	1" X 12"	LF	.68		
	1" X 24"	LF	.78		
ROOF OVERHANG	REMOVE ROOF OVERHANG, INCLUDING SOFFIT, FASCIA, AND GUTTER				
	6"	LF	5.48		
	12"	LF	6.00		
	18"	LF	7.08		
	24"	LF	8.85		
GUTTERS	REMOVE EDGE HUNG GUTTERS FROM BUILDING				
	ALUMINUM, COPPER OR GALVANIZED	LF	.96		
	WOOD	LF	1.34		
DOWN-SPOUTS	REMOVE DOWNSPOUTS FROM BUILDING	LF	.59		
SKYLIGHT	REMOVE GLAZED ROOF SKYLIGHT & CURB UP TO 24 SF	SF	6.00		
METAL PLATFORM AND STEPS	REMOVE EXISTING METAL PLATFORM AND STEPS FROM BUILDING (ONE STORY)	EA	162.00		
WOOD RETAINING WALL	REMOVE 6 X 6 OR 6 X 8 WOOD TIE WALL	SF	.93		

SPECIFICATIONS		UNIT	PRICE	LOCAL AREA MODIFICATION %	LOCAL AREA MODIFICATION PRICE
GARAGE	COMPLETE GARAGE, BUT **NOT** INCLUDING FLOOR SLAB AND DRIVEWAY				
	FRAME	SF	1.47		
	SOLID BRICK	SF	2.33		
	BRICK AND BLOCK	SF	2.00		
	SF = FLOOR AREA				
WOOD PORCH	REMOVE WOOD PORCH UP TO 240 SF. **NO** PATCHING OR REPAIRS INCLUDED				
	ONE STORY	EA	613.50		
	TWO STORY	EA	918.00		
PORCH SCREENING	REMOVE PORCH SCREENING AND INTERMEDIATE POSTS, ETC. INSTALL TEMPORARY SUPPORTS FOR ROOF EA = EACH JOB	EA	174.00		
PORCH FLOORING	REMOVE TONGUE AND GROOVE PORCH FLOORING	SF	.75		
PORCH CEILING	REMOVE WOOD TONGUE & GROOVE PORCH CEILING	SF	.76		
	REMOVE 4x8 1/4" TO 1/2" PLYWOOD PORCH CEILING WITH OR WITHOUT BATTENS	SF	.65		
PORCH RAILING	REMOVE WOOD RAILING AND SPINDLES	LF	.57		
	REMOVE METAL/WROUGHT IRON RAILING AND SPINDLES	LF	.71		
PORCH ROOF	TEAR OFF ENTIRE ROOF STRUCTURE OVER PORCH, UP TO 240 SF, LEAVING ONLY PLATFORM, **NO** POINTING OR PATCHING INCLUDED	EA	408.00		
DECK	REMOVE COMPLETE DECK INCLUDING SURFACE AND FRAMING	SF	1.61		
	REMOVE DECK SURFACE ONLY IN PREPARATION FOR REPLACING	SF	.96		
CHAIN LINK FENCING	REMOVE CHAIN LINK FENCE AND DISPOSE, INCLUDING POSTS SET IN CONCRETE	LF	3.57		
	REMOVE CHAIN LINK FENCE AND STORE FOR RE-USE	LF	2.21		
WOOD FENCING	6'-0" HIGH, 1" X 4" OR 1" X 6" PRIVACY FENCE OR STOCKADE FENCE, INCLUDING POSTS **NOT** SET IN CONCRETE	LF	2.34		
	SAME AS ABOVE WITH POSTS SET IN CONCRETE	LF	3.75		

2. FRAMING AND SHEATHING TEAR-OUT

SPECIFICATIONS		UNIT	PRICE	LOCAL AREA MODIFICATION %	LOCAL AREA MODIFICATION PRICE
BEAM	STEEL I-BEAM OR WIDE FLANGE	LF	7.71		
	WOOD BEAM				
	6 X 8	LF	5.84		
	10 X 12	LF	12.75		
JOISTS	REMOVE WOOD FLOOR JOISTS				
	2 X 6	SF	.65		
	2 X 8	SF	.66		
	2 X 10	SF	.69		
	2 X 12	SF	.72		
SUBFLOOR	REMOVE WOOD SUBFLOOR, 1 X 6 OR 1 X 8	SF	.78		
	REMOVE PLYWOOD SUBFLOOR, NAILED ONLY	SF	.68		
	REMOVE PLYWOOD SUBFLOOR, GLUED AND NAILED	SF	.87		
ROOF SHEATHING	REMOVE SHEATHING				
	1" X 8"	SF	.47		
	1" X 12"	SF	.39		
	1/2" PLYWOOD	SF	.36		
	NOTE: FOR TEARING OUT ROOF COVERING, SEE PAGE 144.				
RAFTERS	REMOVE RAFTERS	SF	.66		
CEILING	REMOVE CEILING JOISTS	SF	.53		
TRUSSES	REMOVE TRUSSES	EA	16.50		

SPECIFICATIONS		UNIT	PRICE	LOCAL AREA MODIFICATION	
				%	PRICE
EXTERIOR DOOR	REMOVE DOOR, FRAME AND TRIM FROM EXISTING EXTERIOR WALL, 3-0 X 6-8	EA	39.30		
	REMOVE WOOD OR ALUMINUM SLIDING GLASS DOOR, FRAME AND TRIM	EA	60.00		
	REMOVE OVERHEAD GARAGE DOOR				
	9 X 7	EA	87.00		
	16 X 7	EA	144.00		
INTERIOR DOOR	REMOVE INTERIOR DOOR, FRAME AND TRIM FROM INTERIOR WALL				
	SINGLE DOOR	EA	19.50		
	DOUBLE DOORS	EA	31.50		
WINDOW	REMOVE WINDOW FROM WALL, INCLUDING TRIM				
	IN FRAME WALL				
	WOOD, TO 15 SF	EA	16.80		
	OVER 15 SF	SF	1.14		
	ALUMINUM, TO 15 SF	EA	10.80		
	OVER 15 SF	SF	.74		
	STEEL CASEMENT, TO 15 SF	EA	33.00		
	OVER 15 SF	SF	2.13		
	IN MASONRY WALL				
	WOOD, TO 15 SF	EA	24.00		
	OVER 15 SF	SF	1.61		
	ALUMINUM, TO 15 SF	EA	17.25		
	OVER 15 SF	SF	1.19		
	STEEL CASEMENT, TO 15 SF	EA	49.50		
	OVER 15 SF	SF	3.30		
	IF WINDOWS HAVE BEEN CAULKED WITH SILICON **ADD**	EA	10.65		

2

2. FLOORING TEAR-OUT

SPECIFICATIONS		UNIT	PRICE	LOCAL AREA MODIFICATION %	LOCAL AREA MODIFICATION PRICE
HARDWOOD FLOOR	REMOVE NAILED HARDWOOD FLOORING 2-1/4" X 25/32"	SF	.65		
	REMOVE HARDWOOD FLOORING IN MASTIC	SF	1.14		
	REMOVE WOOD FLOORING AND SLEEPERS FROM CONCRETE SLAB	SF	.72		
BRICK FLOOR	REMOVE BRICK FLOORING				
	FROM WOOD BASE	SF	.90		
	FROM CONCRETE BASE	SF	2.84		
CERAMIC TILE FLOOR	REMOVE CERAMIC TILE				
	THIN SET	SF	.78		
	MUD SET	SF	1.14		
FLAGSTONE FLOOR	REMOVE SLATE OR FLAGSTONE FLOOR				
	THIN SET	SF	.80		
	MUD SET	SF	1.34		
RESILIENT FLOORING	REMOVE RESILIENT TILE OR SHEET GOODS	SF	.29		
CARPET	REMOVE CARPET				
	BONDED	SF	.44		
	TACKLESS	SF	.14		
UNDER-LAYMENT	REMOVE UNDERLAYMENT, NAILED OR STAPLED	SF	.44		
	NAILED OR STAPLED AND GLUED	SF	.60		

SPECIFICATIONS		UNIT	PRICE	LOCAL AREA MODIFICATION %	LOCAL AREA MODIFICATION PRICE
INTERIOR GUTTING	GUT INTERIOR OF BUILDING LEAVING ONLY LOAD BEARING STUDS, JOISTS, AND CHIMNEY: • REMOVE ALL PLUMBING, HEATING AND ELECTRICAL SYSTEMS • REMOVE INTERIOR WALL COVERINGS EXCEPT PLASTER APPLIED DIRECTLY ON INSIDE OF EXTERIOR MASONRY WALLS • LOAD AND HAUL RUBBISH TO DUMP WITHIN 5 MILES				
	INCLUDING CHUTING RUBBISH TO TRUCK	SF	2.21		
	HAND LOADING INTO TRUCK	SF	2.52		
	SF = TOTAL FLOOR (LIVING) AREA OF BUILDING, INCLUDING BASEMENT				
CEILING COVERING	REMOVE CEILING COVERING FROM CEILING JOISTS				
	PLASTER & GYPSUM LATH	SF	.39		
	PLASTER & WOOD LATH	SF	.54		
	PLASTER & METAL LATH	SF	.87		
	GYPSUM DRYWALL	SF	.29		
	REMOVE WOOD TONGUE & GROOVE CEILING	SF	.78		
CEILING TILE	12" X 12" COMPOSITION TILE				
	GLUED	SF	.60		
	STAPLED	SF	.30		
	24" X 48" TILE IN METAL GRID SYSTEM, INCLUDING GRIDS	SF	.68		
INSULATION	REMOVE INSULATION FROM OPEN WALL OR CEILING				
	LOOSE	SF	.12		
	BATT	SF	.18		

2. INTERIOR WALL TEAR-OUT

2

SPECIFICATIONS		UNIT	PRICE	LOCAL AREA MODIFICATION %	LOCAL AREA MODIFICATION PRICE
COMPLETE INTERIOR WALL	REMOVE NON-BEARING WALL AND COVERINGS, INCLUDING STUDS AND WALL COVERING BOTH SIDES				
	WITH GYPSUM DRYWALL	LF	8.40		
	WITH PLASTER & GYPSUM LATH	LF	9.96		
	WITH PLASTER & WOOD LATH	LF	14.07		
	REMOVE BEARING WALL AND COVERINGS, INCLUDING STUDS AND WALL COVERING BOTH SIDES, INSTALL WOOD HEADER OR STEEL ANGLE SUPPORT AS REQUIRED				
	WITH GYPSUM DRYWALL	LF	21.00		
	WITH PLASTER & GYPSUM LATH	LF	22.50		
	WITH PLASTER & WOOD LATH	LF	27.00		
	SAME AS ABOVE, WITH STEEL BEAM OR FLITCH PLATE SUPPORT				
	WITH GYPSUM DRYWALL	LF	60.90		
	WITH PLASTER & GYPSUM LATH	LF	61.95		
	WITH PLASTER & WOOD LATH	LF	66.00		
HEADER OR STEEL ANGLE	SAME AS ABOVE, INSTALL HEADER OR STEEL ANGLE FLUSH WITH CEILING				
	WITH GYPSUM DRYWALL	LF	34.13		
	WITH PLASTER & GYPSUM LATH	LF	34.88		
	WITH PLASTER & WOOD LATH	LF	36.38		
	SAME AS ABOVE, WITH STEEL BEAM OR FLITCH PLATE FLUSH WITH CEILING				
	WITH GYPSUM DRYWALL	LF	81.00		
	WITH PLASTER & GYPSUM LATH	LF	82.50		
	WITH PLASTER & WOOD LATH	LF	84.00		
COMPLETE METAL STUDWALL	REMOVE METAL STUDWALL FINISHED BOTH SIDES WITH DRYWALL	LF	11.55		
INTERIOR WALL COVERING FROM STUDWALL	GYPSUM DRYWALL	SF	.21		
	GYPSUM LATH & PLASTER	SF	.41		
	METAL LATH & PLASTER	SF	1.01		
	WOOD LATH & PLASTER	SF	.74		
	PLYWOOD PANELING	SF	.21		
	SOLID WOOD PANELING	SF	.48		
	CERAMIC TILE IN MUD	SF	1.19		
	REMOVE PLASTER FROM MASONRY WALL	SF	1.68		

2. MILLWORK TEAR-OUT

SPECIFICATIONS		UNIT	PRICE	LOCAL AREA MODIFICATION %	LOCAL AREA MODIFICATION PRICE
STAIRS	REMOVE SET OF STAIRS, **NO** PATCHING				
	BASEMENT	EA	39.30		
	ATTIC	EA	105.90		
	REMOVE SET OF MAIN STAIRS INCLUDING NEWELS, RAILS AND BALUSTERS, **NO** PATCHING				
	AVERAGE	EA	183.00		
	DIFFICULT	EA	219.00		
	VERY DIFFICULT	EA	273.00		
KITCHEN CABINETS	REMOVE WOOD OR STEEL KITCHEN CABINETS — BASE, WALL OR ISLAND	LF	11.25		
COUNTER-TOP	REMOVE COUNTERTOP FROM BASE CABINET (**NOT** INCLUDING DISCONNECT)	LF	1.92		
BASEBOARD	REMOVE BASEBOARD AND SHOE MOULDING, UP TO 1" X 8" BASEBOARD	LF	.32		
	REMOVE BASEBOARD AND CLEAN AND STORE FOR RE-USE	LF	.45		
MOULDING	REMOVE DOOR OR WINDOW CASING				
	1-PIECE	LF	.50		
	2-PIECE	LF	.63		
	REMOVE BASE, CHAIR RAIL OR CEILING MOULD				
	1-PIECE	LF	.50		
	2-PIECE	LF	.63		
	3-PIECE	LF	.78		
	REMOVE BASE, CHAIR RAIL OR CEILING MOULD, CLEAN AND STORE FOR RE-USE				
	1-PIECE	LF	.56		
	2-PIECE	LF	.92		
	3-PIECE	LF	1.29		

2

2. PLUMBING TEAR-OUT

SPECIFICATIONS		UNIT	PRICE	LOCAL AREA MODIFICATION %	LOCAL AREA MODIFICATION PRICE
KITCHEN PLUMBING	DISCONNECT, REMOVE AND CAP				
	KITCHEN SINK	EA	102.00		
	DISHWASHER	EA	76.50		
	DISPOSAL	EA	33.00		
	BAR SINK	EA	85.50		
	20 GAL. LAUNDRY SINK	EA	67.50		
	DISCONNECT, REMOVE AND CAP ANY TWO OF ABOVE AT SAME TIME **DEDUCT**	EA	10%		
	ANY THREE OR MORE, **DEDUCT**	EA	15%		
BATHROOM PLUMBING	DISCONNECT, REMOVE AND CAP				
	WATER CLOSET	EA	76.50		
	BIDET	EA	102.00		
	STEEL TUB	EA	186.00		
	FIBERGLASS TUB	EA	171.00		
	FIBERGLASS SHOWER/TUB	EA	205.50		
	WHIRLPOOL	EA	256.50		
	WALL-HUNG OR PEDESTAL LAVATORY	EA	76.50		
	19" VANITY AND TOP	EA	90.00		
	60" VANITY AND TOP	EA	118.50		
	DISCONNECT, REMOVE AND CAP ANY TWO OF ABOVE AT SAME TIME **DEDUCT**	EA	10%		
	ANY THREE OR MORE, **DEDUCT**	EA	15%		
	DISCONNECT, REMOVE AND CAP CAST IRON TUB	EA	256.50		
	DISCONNECT, REMOVE AND CAP HOT WATER HEATER				
	30 GALLON	EA	76.50		
	50 GALLON	EA	85.50		
	82 GALLON	EA	93.00		
	REMOVE SHOWER STALL LEAD PAN IN MUD SET TILE	EA	262.50		

2. HEATING AND COOLING TEAR-OUT

2

SPECIFICATIONS		UNIT	PRICE	LOCAL AREA MODIFICATION	
				%	PRICE
HEATING TEAR-OUT	REMOVE BOILER (WITH **NO** ASBESTOS COMPLICATIONS)	EA	738.00		
	REMOVE GAS FURNACE (WITHOUT DUCTWORK)	EA	132.00		
	REMOVE OIL FURNACE (WITHOUT DUCTWORK)	EA	165.00		
	REMOVE BASEBOARD RADIATORS	LF	21.00		
	REMOVE FLUE	LF	2.54		
DRAIN SYSTEM	DRAIN HEATING SYSTEM AND CAP	EA	144.00		
DUCTWORK	TEAR OUT DUCTWORK ONLY				
	4" X 8"	LF	1.44		
	6" X 8"	LF	1.86		
	10" X 12"	LF	2.46		
	6" ROUND	LF	1.31		
	8" FLEX	LF	1.08		
REMOVE SYSTEM	REMOVE ENTIRE BOILER AND RADIATOR SYSTEM				
	ONE STORY	EA	906.00		
	TWO STORY	EA	1,212.00		
	THREE STORY	EA	1,524.00		
	TEAR OUT FURNACE AND ENTIRE DUCTWORK SYSTEM				
	ONE STORY	EA	360.00		
	TWO STORY	EA	513.00		
	THREE STORY	EA	711.00		
WINDOW A/C	REMOVE WINDOW AIR CONDITIONERS				
	SMALL	EA	30.00		
	MEDIUM	EA	51.00		
	LARGE	EA	75.00		

2. ELECTRICAL TEAR-OUT

2

SPECIFICATIONS		UNIT	PRICE	LOCAL AREA MODIFICATION %	LOCAL AREA MODIFICATION PRICE
LIGHT FIXTURES	REMOVE LIGHT FIXTURES:				
	SMALL CEILING FIXTURE	EA	15.00		
	LARGE CEILING FIXTURE	EA	24.00		
	LARGE CHANDELIER	EA	31.50		
	RECESSED FIXTURE	EA	22.50		
	48" FLUORESCENT FIXTURE	EA	18.00		
SWITCHES, OUTLETS	REMOVE SWITCHES AND DUPLEX OUTLETS	EA	9.00		
	REMOVE 220 VOLT OUTLET	EA	12.00		
	REMOVE ATTIC FAN	EA	73.50		
	REMOVE KITCHEN OR BATHROOM EXHAUST FAN	EA	21.00		
	REMOVE HARD WIRED SMOKE ALARM	EA	34.50		
	REMOVE 60 AMP SERVICE	EA	51.00		
	REMOVE 100 AMP SERVICE	EA	61.50		
	REMOVE 150 AMP SERVICE	EA	78.00		
	REMOVE BX CABLE	LF	.86		
	REMOVE ROMEX WIRING	LF	.86		
	REMOVE WHOLE HOUSE CEILING EXHAUST FAN	EA	81.00		
	REMOVE BASEBOARD HEATING SYSTEM	LF	13.13		

SPECIFICATIONS		UNIT	PRICE	LOCAL AREA MODIFICATION %	LOCAL AREA MODIFICATION PRICE
CLEAN & STORE BRICKS	CLEAN BRICKS WITH PNEUMATIC TOOL AND STORE FOR RE-USE IN RESTORATION WORK	EA	.21		
	CLEAN BRICKS BY HAND AND STORE FOR RE-USE IN RESTORATION WORK	EA	.47		
CLEAN & STORE FLAGSTONE OR SLATE	CLEAN FLAGSTONE OR SLATE WITH PNEUMATIC TOOL AND STORE FOR RE-USE IN RESTORATION WORK	EA	.42		
	CLEAN FLAGSTONE OR SLATE BY HAND AND STORE FOR RE-USE IN RESTORATION WORK	EA	.68		
CLEAN WALL OR FLOOR	CLEAN BRICK OR CERAMIC TILE FLOOR OR WALL WITH MURIATIC ACID SOLUTION	SF	.38		
EXTERMI-NATING	EXTERMINATE ROACHES AND VERMIN FROM PRE-MISES EA = 1 TO 3-STORY BUILDING	EA	201.00		
SANDBLAST CLEANING	SANDBLAST EXTERIOR WALL, **NO** REPOINTING OF MORTAR AND **NO** SCAFFOLDING INCLUDED SF = TOTAL AREA, INCLUDING DOOR & WINDOW OPENINGS	SF	1.38		
CLEAN CHIMNEY	TWO STORY HOUSE				
	1 FLUE	EA	94.50		
	2 FLUES	EA	171.00		
	3 FLUES	EA	249.00		

2. SMOKE, FIRE AND FLOOD DAMAGE

2

SPECIFICATIONS		UNIT	PRICE	LOCAL AREA MODIFICATION %	LOCAL AREA MODIFICATION PRICE
FIRE DAMAGE TEAR-OUT	IF ANY TEAR-OUT MATERIALS SHOWN ON THE PRECEDING PAGES ARE FIRE DAMAGED **ADD**	EA	20%		
SCRAPE	SCRAPE FIRE DAMAGED MATERIAL				
	SIDING, BOARDS AND TRIM				
	1 X 6	LF	2.91		
	1 X 8	LF	4.05		
	1 X 10	LF	5.01		
	1 X 12	LF	5.82		
	FRAMING				
	2 X 4	LF	2.94		
	2 X 6	LF	3.59		
	2 X 8	LF	4.62		
	2 X 10	LF	5.36		
	2 X 12	LF	6.47		
	BEAMS				
	4 X 6	LF	4.44		
	6 X 10	LF	6.57		
SMOKE DAMAGE	REMOVE SMOKE FROM WALLS AND CEILINGS	SF	.53		
	SEAL FIRE OR SMOKE DAMAGE				
	AVERAGE	SF	.66		
	EXTENSIVE	SF	1.05		
FLOOD DAMAGE CLEANING	CLEAN UNDAMAGED MATERIAL OF RESIDUE AND MILDEW AND DISINFECT				
	WALLS	SF	1.73		
	FLOOR	SF	1.85		

	SPECIFICATIONS	UNIT	PRICE	LOCAL AREA MODIFICATION %	LOCAL AREA MODIFICATION PRICE
	THE COST OF ASBESTOS ABATEMENT THROUGHOUT THE COUNTRY VARIES CONSIDERABLY. THE DIFFERENCE IN COSTS MAY NOT BE CAUSED AS MUCH BY DIFFERENCES IN LOCAL WAGE RATES AS BY VARYING STATE AND LOCAL GOVERNMENT RULES AND REQUIREMENTS. SOME FACTORS AFFECTING COSTS ARE THE SPECIALIZED EQUIPMENT THAT MAY BE REQUIRED AND THE COST OF DISPOSAL OF THE ASBESTOS-CONTAINING MATERIALS. THE COST OF MOVING THE WASTE TO EPA-APPROVED DUMP SITES WILL VARY ACCORDING TO DISTANCES TRAVELED, DUMPING FEES AND REGULATIONS CONCERNING THE HANDLING OF HAZARDOUS WASTES. BEFORE BIDDING ON AN ASBESTOS ABATEMENT PROJECT, CHECK EPA AND LOCAL REGULATIONS AND OBTAIN FIRM BIDS FOR THE WORK FROM TECHNICIANS WHO YOU ARE CERTAIN ARE PROPERLY QUALIFIED TO COMPLETE THE WORK. THE COSTS SHOWN BELOW ARE INTENDED AS A GENERAL GUIDE, AND ARE **NOT** COSTS FOR ANY PARTICULAR AREA OF THE COUNTRY.				
REMOVAL	REMOVE ASBESTOS FROM CEILING				
	MINIMUM	EA	1,875.00		
		PLUS			
		SF	5.63		
	MAXIMUM	EA	3,735.00		
		PLUS			
		SF	24.38		
	REMOVE ASBESTOS FROM UP TO 6" PIPING				
	MINIMUM	EA	1,875.00		
		PLUS			
		LF	5.63		
	MAXIMUM	EA	3,735.00		
		PLUS			
		LF	42.00		
ENCAPSULATION	ENCAPSULATE ASBESTOS PIPE COVERING WITH SEALANTS				
	MINIMUM	EA	930.00		
		PLUS			
		LF	5.63		
	MAXIMUM	EA	5,625.00		
		PLUS			
		LF	16.80		

3. EXCAVATION AND CLEARING BY HAND

3

SPECIFICATIONS		UNIT	PRICE	LOCAL AREA MODIFICATION %	LOCAL AREA MODIFICATION PRICE
	NOTE: ALL PRICES ARE BASED ON CUBIC **FEET** OF DIRT EXCAVATED				
EXCAVATION	HAND-DIGGING AND PLACING DIRT IN WHEELBARROW OR ON GROUND NEXT TO EXCAVATION. EXCAVATION NOT OVER 6 FEET DEEP. EXCESS DIRT WHEELED 50 FEET OR LESS.				
	DIRT, GRAVEL OR TOPSOIL	CF	1.70		
	HARD CLAY OR LOOSE ROCK	CF	3.05		
FROZEN SOIL	EXCAVATE FROZEN SOIL WITH PICK AND SHOVEL	CF	5.78		
DIFFICULT EXCAVATION	EXCAVATE BY HAND AROUND PIPES, SERVICES AND OTHER OBSTRUCTIONS	CF	4.89		
CRAWL SPACE	DIG OUT EXISTING CRAWL SPACE FOR FULL 18" CLEARANCE UNDER FLOOR JOISTS				
	EASY ACCESS (OPEN FLOOR)	CF	2.33		
	DIFFICULT ACCESS (DIG THROUGH OPEN JOISTS)	CF	4.56		
LOWER EXISTING BASEMENT FLOOR	EXCAVATE BY HAND WITH PNEUMATIC TOOL, WHEEL DIRT TO OUTSIDE OR REMOVE IT BY BELT CONVEYOR. DOES **NOT** INCLUDE BREAK-UP OF EXISTING CONCRETE FLOOR OR UNDERPINNING EXISTING FOUNDATION WALLS				
	EASY ACCESS	CF	2.33		
	DIFFICULT ACCESS	CF	3.87		
	CF = CUBIC FEET OF DIRT EXCAVATED				
CLEAR TREES	CLEAR TREES BY HAND WITH CHAIN SAW, CUT INTO SHORT LENGTHS AND LOAD INTO TRUCK OR STACK				
	8" TREE	EA	70.50		
	12" TREE	EA	84.00		
	18" TREE	EA	109.50		
	24" TREE	EA	115.50		
REMOVE STUMP	REMOVE STUMP FROM GROUND WITH BULLDOZER AND LABORER				
	ADD TO ABOVE	EA	120%		

3. EXCAVATION AND GRADING BY HAND

SPECIFICATIONS		UNIT	PRICE	LOCAL AREA MODIFICATION	
				%	PRICE
BACKFILL BY HAND	BACKFILL BY HAND FROM PILES NEXT TO EXCAVATION	CF	.60		
COMPACT	COMPACT BACKFILLED DIRT				
	WITH HAND TAMPER	CF	.47		
	WITH PNEUMATIC TAMPER	CF	.20		
MOVE FILL DIRT	MOVE AND SPREAD FILL DIRT BY WHEELBARROW AFTER DIRT DUMPED ON PREMISES IN PILES	CF	.47		
FINISH GRADING	PLACING TOPSOIL DELIVERED BY TRUCK, TOPSOIL @ $10 PER CUBIC YARD DELIVERED				
	4"	SF	.36		
	6"	SF	.48		
	RAKING AND SEEDING BY HAND	SF	.30		
	SODDING, LABOR AND MATERIALS BY NURSERY PERSONNEL, 1-1/2" DEEP				
	BLUEGRASS	SF	1.02		
	SUNSHADE	SF	1.14		
AREAWAY	EXCAVATE FOR AND INSTALL CORRUGATED METAL RETAINING WALL AROUND BASEMENT WINDOW, WITH GRAVEL OR LOOSE BRICK LAID OVER 6" SAND BED, GALVANIZED CORRUGATED WALL 24" X 37" NOTE: FOR AREAWAY WITH BRICK WALLS, SEE "MASONRY"	EA	108.00		
SOIL TREATMENT	TREAT SOIL FOR PERMANENT TERMITE PREVENTION, OPEN ACCESS	SF	.33		

3. MACHINE EXCAVATION

3

SPECIFICATIONS		UNIT	PRICE	LOCAL AREA MODIFICATION %	LOCAL AREA MODIFICATION PRICE
	THE PRICES SHOWN BELOW INCLUDE A MACHINE OPERATOR'S OVERHEAD AND PROFIT				
	NOTE: MACHINE DIGGING COSTS HERE ARE BASED ON CUBIC **FEET** OF DIRT. TO CONVERT CUBIC FEET TO CUBIC YARDS, DIVIDE THE TOTAL CUBIC FEET BY 27, AND THE RESULT IS TOTAL CUBIC YARDS.				
EXCAVATION	WITH BACK HOE OR FRONT END LOADER	EA PLUS CF	525.00 .23		
BACKFILL	BACKFILL WITH BULLDOZER OR FRONT END LOADER. **NO** ROUGH GRADING INCLUDED	EA PLUS SF	375.00 .20		
EXCAVATE BY MACHINE & REMOVE EXCESS DIRT	WITH FRONT END LOADER OR BACKHOE AND 2 TRUCKS, INCLUDING 30-MIN. TRIPS PER LOAD FOR DUMPING, DUMPING FEES **NOT** INCLUDED	EA PLUS CF	705.00 .45		
ROOM ADDITION EXCAVATION	EXCAVATE FOR ROOM ADDITION, BASEMENT OR CRAWL SPACE WITH FRONT END LOADER OR BACKHOE, INCLUDING MINIMUM OF 2 FEET BEYOND ADDITION LINE AT ALL WALLS				
	DEPTH BELOW GRADE	EA PLUS	510.00		
	3-0	SF	.33		
	4-0	SF	.42		
	5-0	SF	.54		
	6-0	SF	.65		
	7-0	SF	.74		
	8-0	SF	.78		
	9-6	SF	.98		
	SF = PLAN DIMENSION OF BASEMENT OR GROUND FLOOR				

SPECIFICATIONS		UNIT	PRICE	LOCAL AREA MODIFICATION	
				%	PRICE
FOOTING DRAIN PIPE	4" CORRUGATED PERFORATED PLASTIC PIPE SURROUNDED WITH 6" OF 3/4" AGGREGATE EXTENDING ALONG EXTERIOR PERIMETER OF FOOTING, COVER WITH VAPOR BARRIER, LEVEL GROUND FOR INSTALLATION OF TILE, BUT **NO** DIGGING OUT	LF	4.85		
EXTEND DRAIN PIPE TO OUTLET	DIG AND EXTEND PERIMETER DRAIN PIPE SYSTEM TO "DAYLIGHT" OR A MINIMUM OF 30 FEET TO DRYWELL, SLOPE PIPE 1/8" PER LINEAL FOOT. COVER WITH VAPOR BARRIER, BACKFILL				
	AVERAGE BELOW GRADE				
	12"	LF	6.06		
	24"	LF	7.62		
	36"	LF	9.11		
	48"	LF	11.27		
	60"	LF	13.50		
DRYWELL	CONSTRUCT DRYWELL TO RECEIVE DRAINAGE FROM PERIMETER DRAIN PIPE SYSTEM AS FOLLOWS: DIG AND BURY 55-GALLON CAPACITY STEEL OR PLASTIC DRUM, PERFORATED WITH 2" DIAMETER HOLES AND FILLED WITH 1-1/2" AGGREGATE, COVER WITH VAPOR BARRIER, BACKFILL	EA	504.00		

3. TREATED WOOD RETAINING WALL

3

SPECIFICATIONS		UNIT	PRICE	LOCAL AREA MODIFICATION %	LOCAL AREA MODIFICATION PRICE
WOOD RETAINING WALL (TIE WALL)	• 6" X 8" PRESSURE TREATED TIMBERS • 6" GRAVEL FILL UNDER • 12" GRAVEL BACKFILL BEHIND WALL • 1/2" ROD CONNECTORS, 4'-0" O.C. • TIEBACKS NAILED TO DEADMAN TIMBER (WALL ANCHOR) 6'-0" O.C. EVERY OTHER COURSE *HEIGHT OF WALL ABOVE BOTTOM OF WALL*				
	24"	LF	52.14		
	36"	LF	80.55		
	48"	LF	104.25		
	60"	LF	132.00		
	72"	LF	156.75		
DRAIN PIPE	INSTALL 4" PERFORATED PLASTIC PIPE SURROUNDED WITH 6" OF 3/4" AGGREGATE ALONG BACK OF WALL EXTENDING TO DAYLIGHT OR DRYWELL (SEE "DRYWELL", PAGE 97)	LF	5.25		

4. TYPES OF CONCRETE FOOTINGS, WALLS, PIERS, SLABS

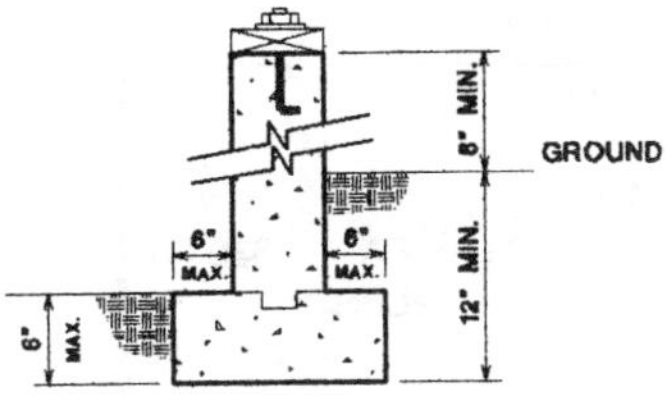

CONCRETE FOUNDATION WALL

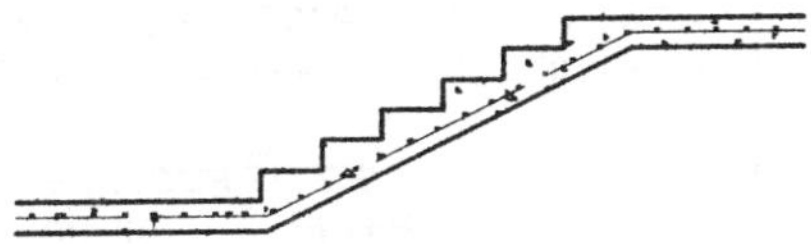

SUSPENDED CONCRETE SLAB & STAIR

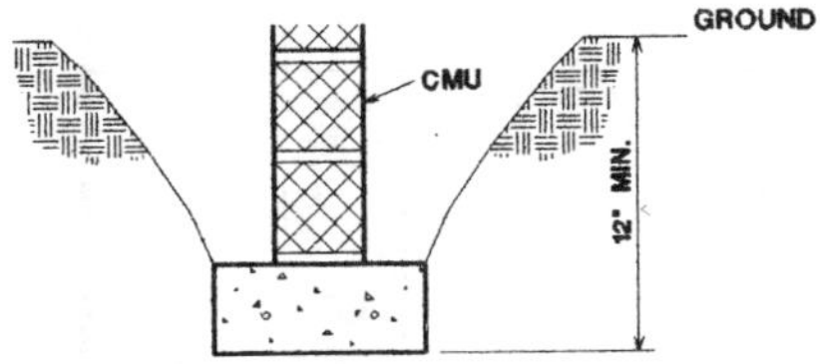

CONCRETE FOOTING FOR BLOCK WALL

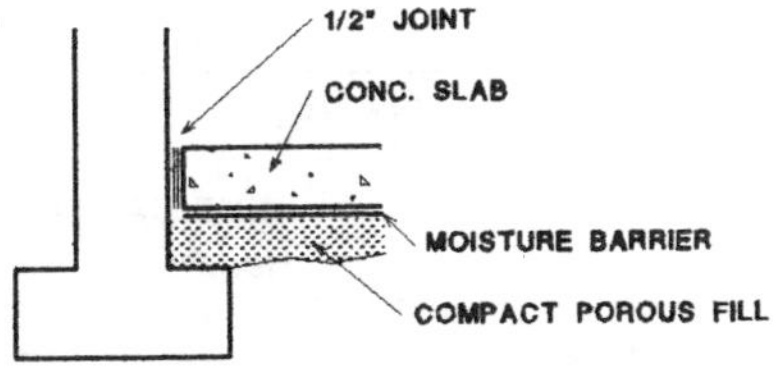

FOOTING/FOUNDATION WITH SLAB

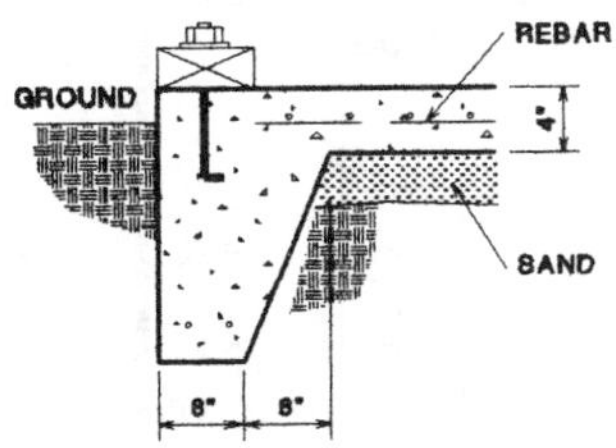

MONOLITHIC FOOTING & SLAB

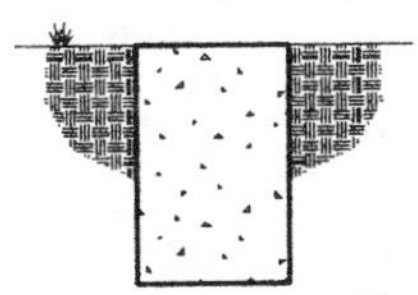

FOOTING TO GRADE

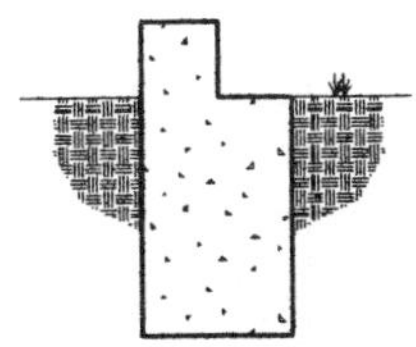

CONCRETE WALL

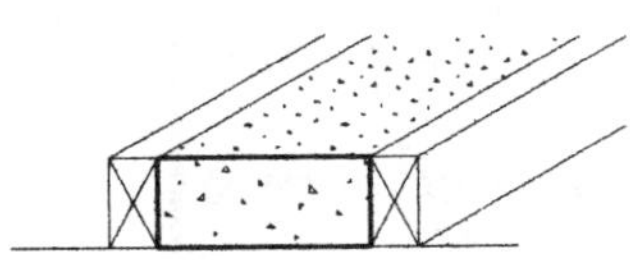

FOUNDATION FOOTING POURED IN FORMS

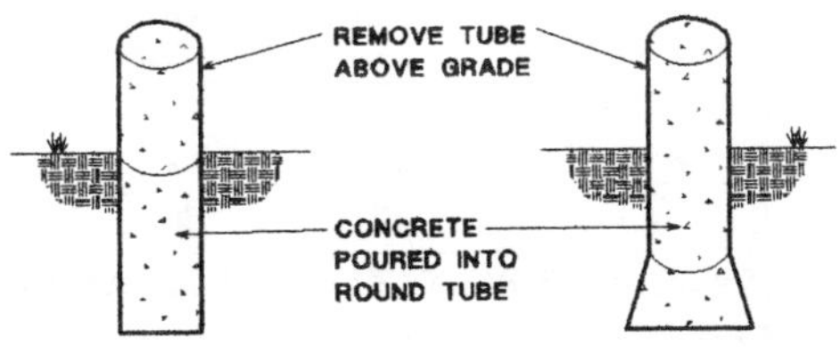

ROUND PIER

ROUND PIER WITH INTEGRAL FOOTING

4. FOOTINGS, MACHINE EXCAVATION

SPECIFICATIONS		UNIT	PRICE	LOCAL AREA MODIFICATION %	LOCAL AREA MODIFICATION PRICE
	THE PRICES SHOWN BELOW FOR MACHINE EXCAVATION INCLUDE A MACHINE OPERATOR'S OVERHEAD AND PROFIT				
4 FOOTINGS FOR WALL	• DIG OUT BY MACHINE, WIDTH OF FOOTING • POUR • BACKFILL AFTER WALL BUILT ON FOOTING • DIG TO 12" BELOW GRADE (BOTTOM OF FOOTING)	EA PLUS	330.00		
	SIZE OF FOOTING				
	Height ***Width***				
	8" 16"	LF	5.66		
	12" 12"	LF	6.33		
	10" 20"	LF	8.84		
	12" 24"	LF	12.69		
	12" 36"	LF	19.10		
	SAME AS ABOVE, DIG OUT TO 24" BELOW GRADE	EA PLUS	330.00		
	SIZE OF FOOTING				
	Height ***Width***				
	8" 16"	LF	6.93		
	12" 12"	LF	7.79		
	10" 20"	LF	10.85		
	12" 24"	LF	15.57		
	12" 36"	LF	23.43		
	SAME AS ABOVE, DIG OUT TO 36" BELOW GRADE	EA PLUS	330.00		
	SIZE OF FOOTING				
	Height ***Width***				
	8" 16"	LF	8.25		
	12" 12"	LF	9.23		
	10" 20"	LF	12.84		
	12" 24"	LF	18.47		
	12" 36"	LF	27.77		
	SAME AS ABOVE, DIG OUT TO 48" BELOW GRADE	EA PLUS	330.00		
	SIZE OF FOOTING				
	Height ***Width***				
	8" 16"	LF	9.54		
	12" 12"	LF	10.67		
	10" 20"	LF	14.85		
	12" 24"	LF	21.35		
	12" 36"	LF	32.10		
	SAME AS ABOVE, DIG OUT TO 60" BELOW GRADE	EA PLUS	330.00		
	SIZE OF FOOTING				
	Height ***Width***				
	8" 16"	LF	10.79		
	12" 12"	LF	12.12		
	10" 20"	LF	16.86		
	12" 24"	LF	24.24		
	12" 36"	LF	36.44		

SPECIFICATIONS		UNIT	PRICE	LOCAL AREA MODIFICATION %	LOCAL AREA MODIFICATION PRICE
FOOTINGS FOR WALL	• DIG OUT BY HAND, WIDTH OF FOOTING • POUR • BACKFILL AFTER WALL BUILT ON FOOTING • DIG TO 12" BELOW GRADE (BOTTOM OF FOOTING) ***SIZE OF FOOTING*** ***Height*** ***Width***				
	8" 16"	LF	9.69		
	12" 12"	LF	9.39		
	10" 20"	LF	13.92		
	12" 24"	LF	18.80		
	12" 36"	LF	28.26		
	SAME AS ABOVE, DIG OUT TO 24" BELOW GRADE ***SIZE OF FOOTING*** ***Height*** ***Width***				
	8" 16"	LF	13.71		
	12" 12"	LF	12.39		
	10" 20"	LF	18.93		
	12" 24"	LF	24.80		
	12" 36"	LF	37.26		
	SAME AS ABOVE, DIG OUT TO 36" BELOW GRADE ***SIZE OF FOOTING*** ***Height*** ***Width***				
	8" 16"	LF	17.70		
	12" 12"	LF	15.39		
	10" 20"	LF	23.91		
	12" 24"	LF	30.80		
	12" 36"	LF	46.26		
	SAME AS ABOVE, DIG OUT TO 48" BELOW GRADE ***SIZE OF FOOTING*** ***Height*** ***Width***				
	8" 16"	LF	21.68		
	12" 12"	LF	18.39		
	10" 20"	LF	28.92		
	12" 24"	LF	36.80		
	12" 36"	LF	55.26		
STEEL BARS IN FOOTING	CONTINUOUS 1/2" STEEL BAR REINFORCED				
	2 BARS	LF	1.80		
	3 BARS	LF	2.64		
	DRILL 1/2" BAR 6" INTO EXISTING FOUNDATION				
	BLOCK	EA	7.10		
	BRICK	EA	7.85		
	CONCRETE	EA	9.20		

4. CONCRETE FOOTINGS TO GRADE

SPECIFICATIONS		UNIT	PRICE	LOCAL AREA MODIFICATION %	LOCAL AREA MODIFICATION PRICE
	THE PRICES SHOWN HERE FOR MACHINE TRENCHING INCLUDE A MACHINE OPERATOR'S OVERHEAD AND PROFIT				
4 MACHINE DUG FOOTINGS	DIG 8" WIDE TRENCH BY MACHINE AND FILL WITH CONCRETE TO GRADE, INCLUDING DIGGING AND POURING	EA PLUS	330.00		
	BOTTOM OF FOOTING UP TO GRADE				
	12"	LF	3.68		
	24"	LF	7.32		
	36"	LF	11.04		
	48"	LF	14.72		
	60"	LF	18.39		
	SAME AS ABOVE, 12" WIDE TRENCH	EA PLUS	330.00		
	BOTTOM OF FOOTING UP TO GRADE				
	12"	LF	5.91		
	24"	LF	11.93		
	36"	LF	17.84		
	48"	LF	23.84		
	60"	LF	29.91		
HAND DUG FOOTINGS	DIG 8" WIDE TRENCH BY HAND AND FILL WITH CONCRETE TO GRADE, INCLUDING DIGGING AND POURING				
	BOTTOM OF FOOTING UP TO GRADE				
	12"	LF	6.21		
	24"	LF	12.36		
	36"	LF	18.59		
	48"	LF	24.78		
	SAME AS ABOVE, 12" WIDE TRENCH				
	BOTTOM OF FOOTING UP TO GRADE				
	12"	LF	8.75		
	24"	LF	17.55		
	36"	LF	26.30		
	48"	LF	35.07		
WALL ABOVE GRADE	• BUILD LEVEL FORMS ABOVE GRADE FOR 6" THICK WALL EXTENDING TO 12" ABOVE GRADE OVER ANY OF ABOVE FOOTINGS • POUR CONCRETE AND STRIP FORMS **ADD**	LF	10.73		
REBAR IN FOOTING	CONTINUOUS NO. 4 (1/2") STEEL BAR REINFORCEMENT				
	2 BARS	LF	1.79		
	3 BARS	LF	2.61		

4. CONCRETE FOOTINGS IN EXISTING EXCAVATION

SPECIFICATIONS		UNIT	PRICE	LOCAL AREA MODIFICATION	
				%	PRICE
WALL FOOTINGS	CONTINUOUS CONCRETE FOOTINGS FOR BLOCK, BRICK OR CONCRETE WALL FOR FULL BASEMENT OR CRAWL SPACE • IN EXISTING EXCAVATION • BUILD LEVEL FORMS FOR FOOTINGS WITH 2 X LUMBER (ONE USE) • POUR CONCRETE • STRIP FORMS				
	SIZE OF FOOTING				
	8" X 16"	LF	15.02		
	10" X 20"	LF	18.53		
	12" X 24"	LF	22.53		
	12" X 36"	LF	25.94		
KEYWAY	PLACE 2" X 4" KEYWAY FOR POURED CONCRETE WALL AND STRIP AFTER POURING	LF	1.73		
STEPPED FOOTING	CONCRETE WALL FOOTINGS BUILT ON INCLINE -- FOR EACH ONE FOOT STEP IN FOOTING, **ADD**				
	8" X 16"	EA	17.15		
	10" X 20"	EA	18.15		
	12" X 24"	EA	19.88		
	12" X 36"	EA	22.50		
	EA = EACH STEP				
REBAR	CONTINUOUS NO. 4 (1/2") STEEL BAR REINFORCEMENT				
	2 BARS	LF	1.79		
	3 BARS	LF	2.61		
COLUMN FOOTING	IN EXISTING BASEMENT OR CRAWL SPACE EXCAVATION, FORM WITH 2 X AND POUR CONCRETE FOOTING FOR STEEL OR MASONRY COLUMN UNDER WOOD OR STEEL BEAM				
	8" THICK X 12" X 12"	EA	24.33		
	12" THICK X 12" X 12"	EA	26.73		
	12" THICK X 16" X 16"	EA	42.45		
	12" THICK X 24" X 24"	EA	57.96		

4. PIER FOOTINGS

4

SPECIFICATIONS					UNIT	PRICE	LOCAL AREA MODIFICATION %	LOCAL AREA MODIFICATION PRICE
PIER FOOTINGS	• DIG OUT BY HAND • POUR CONCRETE • BACKFILL AFTER PIER IS BUILT • BOTTOM OF PIER FOOTING 12" BELOW GRADE							
	SIZE OF PIER FOOTINGS							
		Length	***Width***	***Height***				
		16"	16"	12"	EA	45.00		
		24"	24"	12"	EA	69.00		
	24" BELOW GRADE							
		16"	16"	12"	EA	66.00		
		24"	24"	12"	EA	82.50		
		30"	30"	18"	EA	114.00		
		36"	48"	18"	EA	193.50		
	36" BELOW GRADE							
		16"	16"	12"	EA	70.50		
		24"	24"	12"	EA	87.00		
		30"	30"	18"	EA	124.50		
		36"	48"	18"	EA	234.00		
	48" BELOW GRADE							
		16"	16"	12"	EA	84.00		
		24"	24"	12"	EA	100.50		
		30"	30"	18"	EA	133.50		
		36"	48"	18"	EA	274.50		
	60" BELOW GRADE							
		16"	16"	12"	EA	93.00		
		24"	24"	12"	EA	109.50		
		30"	30"	18"	EA	169.50		
		36"	48"	18"	EA	315.00		

4. ROUND PIERS

4

SPECIFICATIONS		UNIT	PRICE	LOCAL AREA MODIFICATION %	LOCAL AREA MODIFICATION PRICE
ROUND PIERS TO GRADE	• DIG OUT BY HAND • INSTALL ROUND SONOTUBE ON GROUND OR EXISTING FOOTING • POUR CONCRETE • BACKFILL				
	Diameter — ***Bottom of Footing Up To Grade***				
	8" — 12"	EA	25.50		
	24"	EA	42.00		
	36"	EA	58.50		
	48"	EA	73.50		
	60"	EA	93.00		
	12" — 12"	EA	33.00		
	24"	EA	57.00		
	36"	EA	77.25		
	48"	EA	100.50		
	60"	EA	123.00		
	SAME AS ABOVE, INCLUDING CONCRETE INTEGRAL FOOTING UNDER, FLARED TO 16"				
	Diameter — ***Bottom of Footing Up To Grade***				
	8" — 12"	EA	37.50		
	24"	EA	55.50		
	36"	EA	70.50		
	48"	EA	87.00		
	60"	EA	105.00		
	12" — 12"	EA	40.50		
	24"	EA	63.00		
	36"	EA	87.00		
	48"	EA	108.00		
	60"	EA	130.50		
ROUND PIERS EXTENDING ABOVE GRADE	EXTEND ROUND PIERS ABOVE GRADE, CUTTING SONOTUBE OFF AT GRADE AFTER POURING				
	Diameter				
	8"	LF	10.82		
	12"	LF	17.57		
	LF = FROM GRADE TO TOP OF PIER				

4. CONCRETE FOOTING AND FOUNDATION WALL

	SPECIFICATIONS		UNIT	PRICE	LOCAL AREA MODIFICATION %	LOCAL AREA MODIFICATION PRICE
	THE PRICES SHOWN BELOW FOR CONCRETE FOOTINGS AND FOUNDATION WALL INCLUDE A CONCRETE SUBCONTRACTOR'S OVERHEAD AND PROFIT					
CONCRETE FOOTINGS AND FOUNDATION WALL	CONCRETE FOOTINGS AND FOUNDATION WALL FOR FULL BASEMENT OR CRAWL SPACE • FORM CONTINUOUS CONCRETE FOOTINGS TWICE THE THICKNESS OF WALL • POUR FOOTINGS BY CHUTE FROM READY-MIX TRUCK, INCL. KEYWAY • STRIP FORMS AND CLEAN • SET UP PREBUILT SECTIONAL FORMS ON CONCRETE FOOTINGS • SET WINDOWS & STOPS, CUTOUTS & CRAWL HOLES • POUR 3000 PSI CONCRETE MIX INTO FORMS FROM READY-MIX TRUCK • STRIP FORMS AND CLEAN • HAND RUB WALLS WHERE REQUIRED					
COSTS	***WALL THICKNESS***	***WALL HEIGHT ABOVE BOTTOM OF FOOTING***				
	8"	24"	LF	35.72		
		36"	LF	45.60		
		48"	LF	55.49		
		60"	LF	65.37		
		72"	LF	75.26		
		84"	LF	85.14		
		96"	LF	95.03		
	10"	24"	LF	40.56		
		36"	LF	51.20		
		48"	LF	61.83		
		60"	LF	72.47		
		72"	LF	83.10		
		84"	LF	93.74		
		96"	LF	104.37		
	12"	24"	LF	48.81		
		36"	LF	61.59		
		48"	LF	74.37		
		60"	LF	87.15		
		72"	LF	99.93		
		84"	LF	113.61		
		96"	LF	125.49		

4. CONCRETE FOUNDATION WALL

4

SPECIFICATIONS		UNIT	PRICE	LOCAL AREA MODIFICATION %	LOCAL AREA MODIFICATION PRICE
STEEL BAR REINFORCE-MENT	CONTINUOUS 1/2" STEEL BAR REINFORCEMENT IN FOOTINGS 2 BARS 3 BARS LF = FOOTINGS	 LF LF	 1.80 2.61		
	3/8" STEEL BAR REINFORCEMENT IN WALLS, 24" O.C., VERTICAL AND HORIZONTAL GRID SF = **ONE** FACE OF WALL	SF	1.80		
WHEEL CONCRETE	WHEEL CONCRETE FOR CONCRETE FOUNDATION WALL FROM TRUCK UP TO 75 FEET AND POUR INTO FORMS, INCLUDING SETTING UP AND RE-MOVING RAMP IF REQUIRED AND TRANSIT-MIX TRUCK WAITING TIME **ADD**	LF	30%		
PUMP CONCRETE	FOR CONCRETE FOUNDATION WALL, LESS THAN 300' FROM CONCRETE TRUCK, INCLUDING TRUCK AND OPERATOR AND TRANSIT-MIX TRUCK WAIT-ING TIME **ADD**	EA PLUS LF	555.00 12%		

4. MONOLITHIC FOOTING AND SLAB

4

SPECIFICATIONS		UNIT	PRICE	LOCAL AREA MODIFICATION %	LOCAL AREA MODIFICATION PRICE
MONOLITHIC FOOTING AND SLAB (SINGLE POUR)	• DIG, FORM, PLACE REINFORCEMENT • 4" GRAVEL FILL • 6 X 6 #10 WOVEN WIRE MESH • VAPOR BARRIER • POUR AND FINISH CONCRETE • BOTTOM OF SLAB FOOTING 8", TOP OF FOOTING BELOW SLAB 16" WIDE • SLAB THICKNESS 4" • FOOTING BELOW FROST 3 OR 4 SIDES • 1/2" STEEL BARS DRILLED INTO EXISTING BUILDING • TWO 1/2" CONTINUOUS STEEL BARS IN FOOTING				
	TOP SLAB	SF PLUS	4.76		
	FOOTING BELOW GRADE 12"	LF	12.68		
	24"	LF	21.86		
	36"	LF	31.11		
	48"	LF	40.28		
	LF = OUTSIDE PERIMETER MEASUREMENT OF FOOTING				
ADDITIONAL FOOTING	ADDITIONAL 12" WIDE FOOTING UNDER SLAB TO SUPPORT PARTITION, BEAM OR SLAB ITSELF ***BOTTOM OF FOOTING BELOW GRADE***				
	12"	LF	12.68		
	24"	LF	21.86		
WHEEL CONCRETE	WHEEL CONCRETE FOR MONOLITHIC FOOTING AND SLAB FROM TRUCK UP TO 75 FEET AND POUR, INCLUDING SETTING UP AND REMOVING RAMP IF REQUIRED AND TRANSIT-MIX TRUCK WAITING TIME **ADD**	SF	33%		
	SF = TOP SURFACE OF SLAB				
PUMP CONCRETE	FOR MONOLITHIC FOOTING AND SLAB, LESS THAN 300' FROM CONCRETE TRUCK, INCLUDING TRUCK, OPERATOR AND READY-MIX TRUCK WAITING TIME **ADD**	EA PLUS	561.00		
	SF = TOP SURFACE OF SLAB	SF	12%		

SPECIFICATIONS		UNIT	PRICE	LOCAL AREA MODIFICATION %	LOCAL AREA MODIFICATION PRICE
SLAB AND STEP(S) ON GRADE	BUILD 6'-0" x 4'-0" CONCRETE SLAB AND STEP(S) ON GRADE • EXCAVATE TRENCH TO 36" BELOW GRADE • DRILL INTO EXISTING HOUSE WALL AND INSTALL FOUR 1/2" STEEL BARS TO CONNECT FOOTING • POUR CONCRETE FOOTING TO GRADE • BUILD FORMS & POUR CONCR. FOUNDATION WALL ABOVE GRADE TO SUPPORT SLAB & STEP(S) • BUILD FORMS FOR SLAB AND STEP(S) • PLACE RUBBLE OR FORMS IN CENTER • PLACE STEEL • POUR CONCRETE SLAB AND STEP(S) • STRIP & CLEAN FORMS • FINISH CONCRETE • STEPS: 6"-7" RISERS, 12" TREADS	EA	387.00		
	Width of Tread	PLUS			
	48"	Tread	210.00		
	60"	Tread	225.00		
	72"	Tread	234.00		
	EA = EACH JOB TREAD = EACH TREAD				

4. CONCRETE SLAB

4

SPECIFICATIONS		UNIT	PRICE	LOCAL AREA MODIFICATION %	LOCAL AREA MODIFICATION PRICE
BASEMENT FLOOR SLAB	• LEVEL GROUND WITHIN EXISTING FOUNDATION WALLS, BUT **NO** DIGGING OUT • FLOATED & TROWELED • 4" GRAVEL FILL • VAPOR BARRIER UNDER • 6 X 6 #10 WOVEN WIRE MESH				
	3"	SF	4.38		
	4"	SF	4.82		
	5"	SF	5.34		
	6"	SF	5.87		
SLAB, EXTERIOR	ON GRADE, SCREED COAT, FLOATED, 2"	SF	3.41		
	ON GRADE • DIG OUT AND LEVEL GROUND • SPREAD 4" OF SAND, GRAVEL OR CINDERS • FORM AND POUR CONCRETE • FLOATED & TROWELED • VAPOR BARRIER UNDER • 6 X 6 #10 WWM • #4 BARS DRILLED INTO WALL OF EXISTING MASONRY BUILDING				
	4"	SF	4.47		
	5"	SF	4.82		
	6"	SF	5.34		
SLAB, SUSPENDED	SLAB TO BE PLACED ON EXISTING STEEL COLUMN, MASONRY PIERS OR FOUNDATION WALL, NOT OVER 8 FT. ABOVE GRADE • FORM • PLACE STEEL: 1/2" BARS, 12" O.C. • POUR 5"-6" CONCRETE SLAB • STRIP AND CLEAN FORMS FOR RE-USE • FINISH CONCRETE	EA PLUS SF	546.00 8.84		
STEPS, SUSPENDED	SELF-SUPPORTING STEPS TO EXISTING PLATFORM OR AS PART OF SUSPENDED SLAB JOB, SUPPORTED BY EXISTING MASONRY OR UNDISTURBED EARTH • DIG AS REQUIRED • BUILD FORMS • PLACE STEEL AS REQUIRED • POUR CONCRETE STEPS • STRIP AND CLEAN FORMS • FINISH CONCRETE SF = TOTAL TOP SURFACE OF TREADS IN SQ FT	EA PLUS SF	378.00 23.87		

SPECIFICATIONS		UNIT	PRICE	LOCAL AREA MODIFICATION %	LOCAL AREA MODIFICATION PRICE
DRIVEWAY, CONCRETE	• LEVEL BY HAND • FILL WITH UP TO 4" SAND, CINDERS OR GRAVEL • FORM, PLACE #10 WWM AND EXPANSION JOINTS • POUR 4" SLAB, FINISH AND REMOVE FORMS • APRON **NOT** INCLUDED	SF	4.56		
SAND, CINDERS OR GRAVEL FILL UNDER	SAND, CINDER OR GRAVEL FILL UNDER DRIVEWAY, LEVELED BY HAND				
	4"	SF	.72		
	6"	SF	.98		
	8"	SF	1.31		
SIDEWALK, CONCRETE	4" SAND, GRAVEL OR CINDERS BASE, 4" CONCRETE, LEVELING, FORMING, POURING AND FINISHING, REMOVE FORMS	EA PLUS SF	117.00 3.59		
	THE PRICES SHOWN BELOW FOR ASPHALT DRIVEWAYS INCLUDE A SUBCONTRACTOR'S OVERHEAD AND PROFIT				
DRIVEWAY, ASPHALT	TOTAL JOB, INCLUDING FORMING, SPREADING AND ROLLING, OIL BASE, UNDER 250 SF				
	NEW, 4" OVER EXISTING STONE BASE	EA PLUS SF	624.00 2.94		
	REPAVE, 2" ON GOOD BASE	EA PLUS SF	396.00 1.80		
	SAME AS ABOVE, MORE THAN 250 SF				
	4"	EA PLUS SF	1,362.00 2.70		
	2"	EA PLUS SF	852.00 1.56		

5. BLOCK WALL

5

SPECIFICATIONS		UNIT	PRICE	LOCAL AREA MODIFICATION	
				%	PRICE
	ALL MASONRY CONSTRUCTION ON THIS AND THE FOLLOWING PAGE ARE FOR WALLS NOT OVER ONE STORY ABOVE GRADE				
MORE THAN ONE STORY ABOVE GRADE	FOR MASONRY WALLS MORE THAN ONE STORY ABOVE GRADE				
	2 STORIES ABOVE GRADE **ADD**	SF	20%		
	3 STORIES ABOVE GRADE **ADD**	SF	30%		
BLOCK WALL	• ON EXISTING FOOTINGS • TOOLED JOINTS AND SIDES • REINFORCING EVERY SECOND COURSE • STEEL ANGLES AS REQUIRED • TOP COURSE 4" SOLID CAP BLOCK				
	4 X 8 X 16	SF	5.04		
	8 X 8 X 16	SF	6.47		
	12 X 8 X 16	SF	8.70		
	SF = ONE FACE OF WALL				
GLASS BLOCK	• 4" THICK GLASS BLOCK ON EXISTING SUPPORT • TOOLED JOINTS TWO SIDES • CLEAN BOTH SIDES AFTER INSTALLATION				
	6 X 6	SF	46.23		
	8 X 8	SF	36.33		
	12 X 12	SF	39.95		

SPECIFICATIONS		UNIT	PRICE	LOCAL AREA MODIFICATION %	LOCAL AREA MODIFICATION PRICE
BRICK WALL, 4"	• BRICK VENEER WALL ON EXISTING FOOTINGS • OVER EXISTING FRAME WALL • STEEL ANGLES AS REQUIRED • WALL TIES AS REQUIRED • TOOLED JOINTS ONE SIDE • CLEAN ONE SIDE AFTER INSTALLATION	SF	10.38		
SOLID BRICK WALL, 8"	• ON EXISTING FOOTINGS • STEEL ANGLES AS REQUIRED • TOOLED JOINTS ONE SIDE • CLEAN TWO SIDES AFTER INSTALLATION	SF	17.46		
BRICK AND BLOCK WALL	• ON EXISTING FOOTINGS • STEEL ANGLES AS REQUIRED • TOOLED JOINTS ONE SIDE • CLEAN BRICK AFTER INSTALLATION				
	BRICK AND 4" BLOCK	SF	15.65		
	BRICK AND 8" BLOCK	SF	17.01		
REPOINT BRICK WALL	CUT JOINTS IN EXISTING BRICK WALL AND RE-POINT				
	SOFT MORTAR	SF	3.26		
	HARD MORTAR	SF	6.72		
PATCH MASONRY	REMOVE DEFECTIVE MATERIALS AND PATCH EXISTING WALL, TOOTHING IN TO EXISTING WALL				
	BRICK	SF	17.21		
	BLOCK	SF	12.65		

5. PIERS

SPECIFICATIONS		UNIT	PRICE	LOCAL AREA MODIFICATION %	LOCAL AREA MODIFICATION PRICE
BLOCK PIER	• ON EXISTING FOOTING • TOOLED JOINTS • TOP 4" SOLID BLOCK OR CONCRETE FILLED				
	16" X 16"	LF UP	22.13		
	24" X 24"	LF UP	28.67		
	LF UP = TOP OF FOOTING TO TOP OF PIER				
BRICK PIER	• ON EXISTING FOOTING • TOOLED JOINTS • CLEAN BRICK AFTER INSTALLATION				
	12" X 12"	LF UP	35.49		
	16" X 16"	LF UP	45.75		
	LF UP = TOP OF FOOTING TO TOP OF PIER				

5. PLATFORM AND STEPS WITH BRICK & BLOCK BUILD-UP

SPECIFICATIONS		UNIT	PRICE	LOCAL AREA MODIFICATION %	LOCAL AREA MODIFICATION PRICE
SLAB AND STEP(S) ON GRADE	BUILD 6'-0" x 4'-0" CONCRETE SLAB AND STEP(S) ON GRADE • EXCAVATE TRENCH TO 36" BELOW GRADE • DRILL INTO EXISTING HOUSE WALL AND INSTALL FOUR 1/2" STEEL BARS TO CONNECT FOOTING • POUR CONCRETE FOOTING TO GRADE • BUILD BRICK & BLOCK FOUNDATION WALL ABOVE GRADE TO SUPPORT SLAB & STEP(S) • BUILD FORMS FOR SLAB AND STEP(S) • PLACE RUBBLE OR FORMS IN CENTER • PLACE STEEL • POUR CONCRETE SLAB AND STEP(S) • STRIP AND CLEAN FORMS • FINISH CONCRETE • STEPS: 6"-7" RISERS, 12" TREADS	EA	381.00		
	Width of Tread	PLUS			
	48"	Tread	240.00		
	60"	Tread	255.00		
	72"	Tread	264.00		
	EA = EACH JOB TREAD = EACH TREAD				

5

5. CHIMNEY

SPECIFICATIONS		UNIT	PRICE	LOCAL AREA MODIFICATION %	LOCAL AREA MODIFICATION PRICE
CHIMNEY	• BRICK • CEMENT WASH AT TOP • ON EXISTING FOUNDATION • CLAY TILE FLUE LINERS				
	WITH ONE 8" X 8" FLUE	LF	80.27		
	WITH TWO 8" X 8" FLUES	LF	92.91		
	WITH ONE 8" X 12" FLUE	LF	104.94		
	WITH TWO 8" X 12" FLUES	LF	132.24		
	LF = LF UP FROM FOUNDATION TO TOP OF CHIMNEY				
INSTALL FLUE(S) IN EXISTING CHIMNEY	TEAR OUT PORTION OF CHIMNEY FROM INSIDE HOUSE AND INSTALL CLAY TILE FLUE LINING(S), REPLACE BRICKWORK, **NO** PLASTER PATCHING				
	ONE FLUE	LF	59.85		
	TWO FLUES	LF	77.22		
	SAME AS ABOVE, WORKING FROM OUTSIDE ON SCAFFOLDING				
	ONE FLUE	LF	56.22		
	TWO FLUES	LF	72.03		
POURED CONCRETE FLUE LINER IN EXISTING CHIMNEY	STRAIGHT CHIMNEY, PORTLAND CEMENT AND INSULATING AGGREGATE POURED AROUND TEMPORARY INFLATED RUBBER FORM OR STEEL LINER WITH 1/16" GLAZED FINISH ON INSIDE OF NEW FLUE LINER	EA PLUS LF	939.00 57.00		
	FOR EACH ANGLE IN CHIMNEY **ADD**	EA	324.00		
	CLEANOUT DOOR EA = EACH OPENING	EA	108.00		
	TEAR OUT OLD FLUE LINER	LF	21.75		

SPECIFICATIONS		UNIT	PRICE	LOCAL AREA MODIFICATION	
				%	PRICE
FIREPLACE	• BRICK WITH FIREBRICK LINED FIRE HEARTH • DAMPER AND CLEANOUT • BRICK, SLATE OR TILE FRONT HEARTH • SIZE: WIDTH 36" HEIGHT 29" DEPTH 16"	EA	3,642.00		
REPLACE LINTEL OVER EXISTING FIREPLACE	CUT AWAY BRICKWORK AND REMOVE EXISTING LINTEL, INSTALL NEW LINTEL AND REPLACE BRICKWORK 48" LINTEL	EA	486.00		
	72" LINTEL	EA	642.00		
RAISED HEARTH	HEARTH RAISED UP TO ONE FOOT ABOVE FLOOR **ADD**	EA	177.00		
DAMPER IN EXISTING FIREPLACE	REMOVE PORTION OF CHIMNEY ABOVE FIRE-PLACE AND INSTALL NEW DAMPER, REPLACE BRICKWORK	EA	630.00		
ALTER EXISTING FIREPLACE	BREAK THROUGH INTO EXISTING BRICKED UP FIREPLACE, CLEAN OUT AND REPAIR FOR USE WITH EXISTING DAMPER	EA	342.00		
	BREAK THROUGH BACK WALL OF EXISTING FIRE-PLACE, INSTALL NEW FIREPLACE OPENING FOR BACK-TO-BACK FIREPLACES WITH SAME FLUE AND OPEN FROM ROOM TO ROOM	EA	1,791.00		
	GLASS DOORS INSTALLED ON ONE FIREPLACE OPENING ABOVE	EA	417.00		
REBUILD FIREBOX	REBUILD EXISTING FIREBOX WITH NEW FIREBOX APPROXIMATELY 42" X 21"	EA	486.00		
REBUILD FIREPLACE	REBUILD FIREBOX, INSTALL THROAT-TYPE DAMP-ER, REBUILD FIREPLACE UP TO 56" ABOVE FLOOR	EA	1,641.00		

5. UNDERPINNING

SPECIFICATIONS		UNIT	PRICE	LOCAL AREA MODIFICATION %	LOCAL AREA MODIFICATION PRICE
UNDERPIN EXISTING FOUNDATION WALL	EXCAVATE AS REQUIRED UNDER EXISTING FOOTING IN ALTERNATE 4-FOOT SECTIONS POUR REINFORCED CONCRETE UNDER AND BUILD NEW BLOCK WALL FROM TOP OF NEW FOOTING TO BOTTOM OF OLD FOOTING BREAK OFF EXISTING FOOTING PROJECTION				
	Below Existing Footing				
	24"	LF	209.04		
	48"	LF	227.76		
	72"	LF	277.34		
UNDERPIN CORNER OF EXISTING BRICK HOUSE (EXTERIOR)	EXCAVATE AS REQUIRED UNDER EXISTING CORNER AND POUR REINFORCED CONCRETE UNDER EXISTING FOOTING OR FOUNDATION WALL				
	Below Grade				
	48"	EA	639.00		
	72"	EA	819.00		
	96"	EA	1,299.00		

SPECIFICATIONS		UNIT	PRICE	LOCAL AREA MODIFICATION %	LOCAL AREA MODIFICATION PRICE
RETAINING WALL	• CONCRETE FOOTING • SOLID MASONRY WALL WITH BRICK FACING • 4" TILE WEEPHOLES 10 FT. O.C. WITH 1 CU.FT. CRUSHED STONE OR GRAVEL AT EACH WEEP-HOLE • PARGE BACK WALL WITH 1/2" CEMENT PLAS-TER OR 2 COATS HOT TAR • REINFORCING RODS AS REQUIRED				
	Height of Wall Above Bottom of Footing				
	40"	LF	84.09		
	48"	LF	88.79		
	56"	LF	104.33		
	64"	LF	122.90		
	72"	LF	135.53		
	80"	LF	172.64		
DRAIN PIPE	INSTALL 4" PERFORATED PLASTIC PIPE SUR-ROUNDED WITH 6" OF 3/4" AGGREGATE ALONG BACK OF FOOTING EXTENDING TO DAYLIGHT OR DRYWELL (SEE PAGE 97)	LF	4.55		

5. PATIO

5

SPECIFICATIONS		UNIT	PRICE	LOCAL AREA MODIFICATION	
				%	PRICE
PATIO, BRICK	ON GROUND IN SAND WITH SAND IN JOINTS, BRICKS LAID FLAT, INCLUDING EXCAVATION	SF	14.70		
	IN SAND OVER EXISTING CONCRETE SLAB, SAND IN JOINTS	SF	10.95		
	IN CONCRETE BED WITH CONCRETE JOINTS OVER EXISTING CONCRETE SLAB	SF	12.98		
	BRICK LAID ON EDGE IN CONCRETE BED WITH CONCRETE JOINTS OVER EXISTING CONCRETE SLAB	SF	24.60		
PATIO, FLAGSTONE	ON GROUND IN SAND WITH SAND IN JOINTS, INCLUDING EXCAVATION, 3/4" - 1"				
	IRREGULAR PA.	SF	10.58		
	REGULAR PA.	SF	12.08		
	IRREGULAR VT.	SF	11.25		
	REGULAR VT.	SF	20.40		
	REGULAR TENN.	SF	20.40		
	IN CONCRETE BED WITH CONCRETE JOINTS OVER EXISTING CONCRETE SLAB				
	IRREGULAR PA.	SF	10.50		
	REGULAR PA.	SF	11.40		
	IRREGULAR VT.	SF	10.95		
	REGULAR VT.	SF	20.03		
	REGULAR TENN.	SF	20.03		

	SPECIFICATIONS	UNIT	PRICE	LOCAL AREA MODIFICATION %	LOCAL AREA MODIFICATION PRICE
	PRICES SHOWN BELOW FOR STUCCO WORK INCLUDE A SUB-CONTRACTOR'S OVERHEAD AND PROFIT				
STUCCO	CEMENT SAND FINISH STUCCO ON CLEAN MASONRY WALL, TWO 1/2" COATS	SF	6.35		
	CEMENT SAND FINISH STUCCO OVER EXISTING SHEATHING • #15 FELT PAPER • SELF FURRING WIRE MESH • OUTSIDE CORNER BEADS AS NECESSARY • METAL LATH STRIPS AROUND WINDOWS AND DOORS • SCRATCH COAT, BROWN COAT AND FINISH COAT	SF	8.27		
	E.I.F.S. (EXTERIOR INSULATION & FINISH SYSTEM) 1/8" FLEXIBLE STUCCO OVER 1" FOAM SHEATHING (INCLUDED)	SF	9.60		
PATCH	REMOVE DEFECTIVE MATERIALS AND CLEAN AND PREPARE AREA TO BE PATCHED, 2 OR 3 COATS STUCCO				
	3 COATS ON FRAMING	SF	12.00		
	2 COATS ON MASONRY	SF	7.58		
WATER-PROOF PAINT	FILL CRACKS OVER 1/8" WIDE AND APPLY 2 COATS WATERPROOFING PAINT OVER MASONRY WALL	SF	.66		
SILICON	CLEAR SILICON OVER EXISTING MASONRY WALL, 1 COAT SPRAYED	SF	.36		
PARGING	CEMENT PARGING, 1/2" THICK, TWO COATS	SF	1.80		
ASPHALT COATING	ASPHALT COATING APPLIED TO MASONRY WALL				
	1 COAT	SF	.75		
	2 COATS	SF	1.08		
	TWO LAYERS TARRED PAPER COVERED WITH HOT TAR OR ASPHALT ON MASONRY WALL	SF	1.50		

6. ALL WEATHER WOOD FOUNDATION

SPECIFICATIONS		UNIT	PRICE	LOCAL AREA MODIFICATION	
				%	PRICE
TREATED WOOD FOUNDATION	• 2" X 12" PRESSURE TREATED WOOD PLATE ON LAYER OF STONE OR GRAVEL • LAYER OF GRAVEL UP TO 3/4" OR CRUSHED STONE UP TO 1/2" EXTENDING MINIMUM 6" BEYOND PLATE AND 4" DEEP • CDX TREATED PANELS • CORROSION RESISTANT NAILS AND FASTENERS • BATT INSULATION BETWEEN STUDS • 15# FELT PAPER • 6-MIL POLYETHYLENE				
	Stud Size — ***CDX Panel Size***				
	2" X 4" — 1/2"	SF	7.52		
	2" X 6" — 1/2"	SF	9.92		
	2" X 8" — 3/4"	SF	11.42		
	SF = ONE FACE OF WALL				

6

SPECIFICATIONS		UNIT	PRICE	LOCAL AREA MODIFICATION	
				%	PRICE
STEEL BEAM	STEEL BEAM BEARING ON EXISTING MASONRY OR STEEL COLUMN SUPPORTS, INCLUDING BEARING PLATES				
	8" #13	LF	32.93		
	8" #17	LF	40.89		
WOOD BEAM	WOOD BEAM BEARING ON EXISTING SUPPORTS, BEAM EITHER SOLID OR BUILT-UP, PRESSURE TREATED LUMBER				
	4 X 8	LF	6.21		
	4 X 10	LF	10.67		
	6 X 8	LF	11.63		
	6 X 10	LF	17.94		
LAMINATED WOOD BEAM	LAMINATED VENEER LUMBER ON EXISTING SUPPORTS				
	4" X 9-1/4"	LF	28.50		
	4" X 11-1/4"	LF	33.00		
FLITCH PLATE, STEEL	3/8" X 9" STEEL PLATE BETWEEN TWO 2x10s ON EXISTING SUPPORTS, INCLUDING PLATE, LUMBER, NUTS, BOLTS	LF	29.40		
FLITCH PLATE, WOOD	1/2" X 9" PLYWOOD GLUED AND BOLTED BETWEEN TWO 2 X 10s ON EXISTING SUPPORTS, INCLUDING PLYWOOD, LUMBER, GLUE, NUTS, BOLTS	LF	14.37		
STEEL COLUMN	3" HOLLOW STEEL COLUMN WITH BASE AND CAP UP TO 8'-0"	EA	78.00		
	3" HOLLOW ADJUSTABLE STEEL COLUMN WITH BASE AND CAP UP TO 8'-0"	EA	68.25		

6. FLOOR FRAMING

SPECIFICATIONS		UNIT	PRICE	LOCAL AREA MODIFICATION %	LOCAL AREA MODIFICATION PRICE
SLEEPERS ON EXISTING DECK	ON EXISTING WOOD, CONCRETE OR MASONRY DECK • SOLID BRIDGING • JOIST HEADERS AS REQ'D • VAPOR BARRIER • NAILS • SLEEPER JOISTS 16" O.C.				
	2" X 3"	SF	1.77		
	2" X 4"	SF	1.95		
	2" X 6"	SF	2.24		
	SAME AS ABOVE, USING PRESSURE-TREATED LUMBER				
	2" X 3"	SF	1.86		
	2" X 4"	SF	2.04		
	2" X 6"	SF	2.40		
	IF SLEEPERS ARE RIPPED LENGTHWISE TO FIT **ADD**	SF	.41		
	NOTE: FLOOR JOISTS PRICES BELOW ARE BASED ON USING JOISTS TO FULL PERMISSIBLE LENGTHS				
MUDSILL	• PRESSURE-TREATED DIMENSION FIR OR PINE • SILL PLATE • ANCHOR BOLTS, 4' OC				
	2" X 4"	LF	2.45		
	2" X 6"	LF	2.85		
	2" X 8"	LF	3.26		
JOISTS, FIRST FLOOR	• FRAMING LUMBER • BRIDGING • JOIST HANGERS • LAG BOLTS • NAILS • JOIST HEADERS AS REQ. • JOISTS 16" O.C.				
	2" X 6"	SF	2.45		
	2" X 8"	SF	3.18		
	2" X 10"	SF	3.77		
	2" X 12"	SF	4.26		
JOISTS, SECOND FLOOR	• FRAMING LUMBER • BRIDGING • JOIST HEADERS AS REQUIRED • JOIST HANGERS • LAG BOLTS • NAILS • JOISTS 16" O.C.				
	2" X 6"	SF	2.54		
	2" X 8"	SF	3.27		
	2" X 10"	SF	3.86		
	2" X 12"	SF	4.35		

SPECIFICATIONS		UNIT	PRICE	LOCAL AREA MODIFICATION %	LOCAL AREA MODIFICATION PRICE
WOOD I-JOISTS	PREFABRICATED I-SHAPED JOISTS, 16" O.C. • 3/8" WEB • 1-5/8" TOP AND BOTTOM CHORDS • NAILS • JOIST HEADERS • JOIST HANGERS • **NO** BRIDGING				
	9-1/4"	SF	3.41		
	12"	SF	4.08		
SUBFLOOR	PLYWOOD CDX, GLUED AND NAILED TO JOISTS, INCLUDING ALL MATERIALS AS REQUIRED				
	3/8"	SF	1.20		
	1/2"	SF	1.31		
	5/8"	SF	1.38		
	PLYWOOD, GLUED AND NAILED, INCLUDING ALL MATERIALS AS REQUIRED				
	3/4"	SF	1.49		
	T & G 5/8"	SF	1.77		
	T & G 3/4"	SF	1.94		
	3/4" PINE, FIR OR HEMLOCK, NAILED				
	HORIZONTAL	SF	1.71		
	DIAGONAL	SF	2.16		
	ORIENTED STRAND BOARD				
	1/2"	SF	1.38		
	5/8"	SF	1.47		
	3/4"	SF	1.61		

7. WALL FRAMING AND FURRING

7

SPECIFICATIONS		UNIT	PRICE	LOCAL AREA MODIFICATION %	LOCAL AREA MODIFICATION PRICE
WOOD STUDS	BEARING WALL, EXTERIOR OR INTERIOR • SOLE PLATE • TWO 2" X 4" OR 2" X 6" CAP • AVERAGE NUMBER OF HEADERS FOR DOORS AND WINDOWS • 16" O.C.				
	2" X 4"	SF	2.07		
	2" X 6"	SF	2.64		
	SF = TOTAL WALL AREA				
	NON-BEARING WALL • SOLE PLATE • ONE CAP • AVERAGE AMOUNT OF FRAMING FOR DOORS, CLOSETS AND CORNERS • 16" O.C.				
	2" X 3"	SF	1.64		
	2" X 4"	SF	2.07		
	2" X 6"	SF	2.60		
STEEL STUDS	20 GA. STEEL STUDS AND TRACK				
	1-1/2"	SF	1.17		
	2-1/2"	SF	1.28		
	3-1/2"	SF	1.38		
	5-1/2"	SF	1.61		
DIFFICULT WALL FRAMING	FOR DIFFICULT WALL FRAMING, CUT UP WITH BAY AND BOW WINDOWS, MANY CORNERS, ETC. **ADD**	SF	30%		
KNEEWALL	2 X 4 KNEEWALL SUPPORTING RAFTERS • SOLE PLATE • ONE CAP • STUDS CUT AT ANGLE TO FIT RAFTERS				
	Kneewall Height				
	4'-0"	LF	9.92		
	6'-0"	LF	12.44		
	8'-0"	LF	15.20		
FURRING	OVER FRAMING, 16" O.C.				
	1" X 2"	SF	.60		
	1" X 3"	SF	.69		
	OVER STRAIGHT AND PLUMB MASONRY WALLS, 16" O.C.				
	1" X 2"	SF	.83		
	1" X 3"	SF	.92		
	OVER CROOKED MASONRY WALLS, INCLUDING SHIMS, 16" O.C.				
	1" X 2"	SF	1.07		
	1" X 3"	SF	1.16		

SPECIFICATIONS		UNIT	PRICE	LOCAL AREA MODIFICATION	
				%	PRICE
SHEATHING	PLYWOOD CDX				
	3/8"	SF	1.19		
	1/2"	SF	1.31		
	COMPOSITION SHEATHING, INCLUDING 1/2" PLY-WOOD OR 1 X 6 DIAGONAL BRACING AT EACH CORNER, 1/2" X 4 X 8 SHEETS				
	ASPHALT IMPREGNATED	SF	1.04		
	SAME AS ABOVE, USING FOIL-FACED FOAM SHEATHING				
	1/4"	SF	1.05		
	1/2"	SF	1.14		
	5/8"	SF	1.19		
	3/4"	SF	1.29		
	SHEATHING PINE, 3/4"				
	HORIZONTAL	SF	1.77		
	DIAGONAL	SF	2.28		
	ORIENTED STRAND BOARD				
	3/8"	SF	1.29		
	1/2"	SF	1.38		

7

8. ROOF FRAMING

SPECIFICATIONS		UNIT	PRICE	LOCAL AREA MODIFICATION %	LOCAL AREA MODIFICATION PRICE
	MEASURING SQUARE FOOTAGE OF NEW RAFTERS AND SHEATHING TO OBTAIN SQUARE FOOTAGE OF NEW ROOF RAFTERS AND SHEATHING, MEASURE THE ACTUAL LENGTH OF RAFTERS FROM RIDGEBOARD TO EAVE AND MULTIPLY BY THE OPPOSITE DIMENSION (WIDTH OF THE ROOF). BE SURE TO INCLUDE OVERHANGS AND ANY PORTION OF THE NEW ROOF WHICH WILL OVERLAY THE EXISTING ROOF.				
FLAT ROOF OR SHED ROOF	• RAFTERS 16" O.C. • BRIDGING • NAILS • FRAMING FOR OVERHANG AND EAVES				
	2" X 6"	SF	2.04		
	2" X 8"	SF	2.79		
	2" X 10"	SF	3.68		
	2" X 12"	SF	4.58		
CEILING JOISTS	ADD CEILING JOISTS TO SHED ROOF FRAMING				
	2" X 6" **ADD**	SF	1.74		
	2" X 8" **ADD**	SF	2.37		
	2" X 10" **ADD**	SF	2.79		
	SF = AREA COVERED BY CEILING JOISTS, **NOT** ROOF AREA				
GABLE ROOF	• RIDGE BOARD • RAFTERS 16" O.C. • CEILING JOISTS 16" O.C. • COLLAR TIES, 1" X 6", 48" O.C. • NAILS • FRAMING FOR OVERHANG AND EAVES				
	Rafters / ***Ceiling Joists***				
	2" X 6" / 2" X 6"	SF	3.83		
	2" X 8" / 2" X 6"	SF	4.53		
	2" X 10" / 2" X 8"	SF	5.94		
	2" X 12" / 2" X 10"	SF	7.20		
OMIT CEILING JOISTS	OMIT CEILING JOISTS				
	2" X 6" **DEDUCT**	SF	1.74		
	2" X 8" **DEDUCT**	SF	2.37		
	2" X 10" **DEDUCT**	SF	2.79		
	SF = AREA COVERED BY CEILING JOISTS, **NOT** ROOF AREA				
OMIT COLLAR TIES	OMIT COLLAR TIES, 1" X 6", 48" O.C. **DEDUCT**	SF	.09		
	SF = ROOF AREA				
DIFFICULT ROOF FRAMING	DIFFICULT ROOF FRAMING, CUT UP WITH VALLEYS AND DORMERS				
	SOME DIFFICULTY **ADD**	SF	.45		
	DIFFICULT **ADD**	SF	.95		
	VERY DIFFICULT **ADD**	SF	1.50		

8

SPECIFICATIONS		UNIT	PRICE	LOCAL AREA MODIFICATION %	LOCAL AREA MODIFICATION PRICE
GAMBREL ROOF	• RIDGEBOARD • HEADERS • RAFTERS 16" O.C. • 2" X 6" CEILING JOISTS 16" O.C. • 2" X 4" STUDWALL 16" O.C. • NAILS • FRAMING FOR CHIMNEY, EAVES • 12" OVERHANG • SLOPE OF UPPER RAFTERS 7/12				
	2" X 6"	SF	5.61		
	2" X 8"	SF	6.42		
HIP ROOF	• RIDGEBOARD • HEADERS • HIP AND JACK RAFTERS 16" O.C. • CEILING JOISTS 16" O.C. • FRAMING FOR CHIMNEY, EAVES • 12" OVERHANG				
	Rafters — ***Ceiling Joists***				
	2" X 6" — 2" X 6"	SF	4.50		
	2" X 8" — 2" X 6"	SF	5.52		
	2" X 10" — 2" X 8"	SF	7.14		
DIFFICULT ROOF FRAMING	DIFFICULT ROOF FRAMING, CUT UP WITH VALLEYS AND DORMERS				
	SOME DIFFICULTY **ADD**	SF	.45		
	DIFFICULT **ADD**	SF	.95		
	VERY DIFFICULT **ADD**	SF	1.43		
TRUSS ROOF	• SHOP BUILT BY OTHERS AND DELIVERED TO JOB • PLACED BY CRANE OR BY HAND • METAL PLATE CONNECTORS • 2" X 4" TOP AND BOTTOM CHORDS • INCLUDES 2 GABLE ENDS 24" O.C.				
	16" O.C.	SF	6.42		
	24" O.C.	SF	4.47		

8. ROOF FRAMING, SHEATHING

8

SPECIFICATIONS		UNIT	PRICE	LOCAL AREA MODIFICATION	
				%	PRICE
PLYWOOD	CDX PLYWOOD				
	3/8"	SF	1.14		
	1/2"	SF	1.23		
	5/8"	SF	1.32		
	3/4"	SF	1.43		
	ORIENTED STRAND BOARD				
	3/8"	SF	1.28		
	1/2"	SF	1.37		
	5/8"	SF	1.50		
STRIP SHEATHING	STRIP (SPACED) SHEATHING (UNDER WOOD SHAKES OR SHINGLES)				
	1" X 3", 5-1/2" O.C.	SF	1.13		
	1" X 3", 7-1/2" O.C.	SF	.99		
	1" X 4", 7-1/2" O.C.	SF	1.17		
CATHEDRAL CEILING, EXPOSED BEAMS	• GABLE ROOF • 4" X 8" RAFTERS, 48" O.C. • RIDGEBOARD • NAILS AND FASTENERS • FRAMING FOR OVERHANG AND EAVES • COVERED WITH 2" X 6" T&G SHEATHING • #2 CONSTRUCTION GRADE LUMBER SF = ROOF SURFACE AREA	SF	17.61		
	2" RIGID FOAM INSULATION AND 1/2" CDX PLYWOOD INSTALLED OVER ABOVE **ADD**	SF	4.95		
GABLE (DOGHOUSE) DORMER	GABLE (DOGHOUSE) DORMER BUILT AT TIME OF CONSTRUCTING ANY OF ABOVE ROOFS, INCLUDING RAFTERS, RIDGEBOARD, CEILING JOISTS, STUDS, PLATES, SHEATHING SF = DORMER FLOOR AREA	SF	11.55		
SOFFIT NAILER	2" X 4" FRAMING TO SUPPORT SOFFIT				
	Soffit Width				
	12"	LF	1.80		
	18"	LF	2.09		
	24"	LF	2.36		

Measuring Square Footage of Roof

New roofs and re-roofing are measured the same way. To obtain the square footage of roof, measure from ridge to eave and multiply by the opposite dimension (the width of the roof). Be sure to include overhangs and any portion of the roof which overlays the existing roof.

In measuring for a re-roofing job, it is not always possible to measure the exact dimensions of the roof. In this case, you may estimate the number of inches exposure in each course of shingles, count the courses, and multiply that total by the number of inches per course. Convert that figure to feet and multiply by the width of the building (which will usually be approximately the same as the roof).

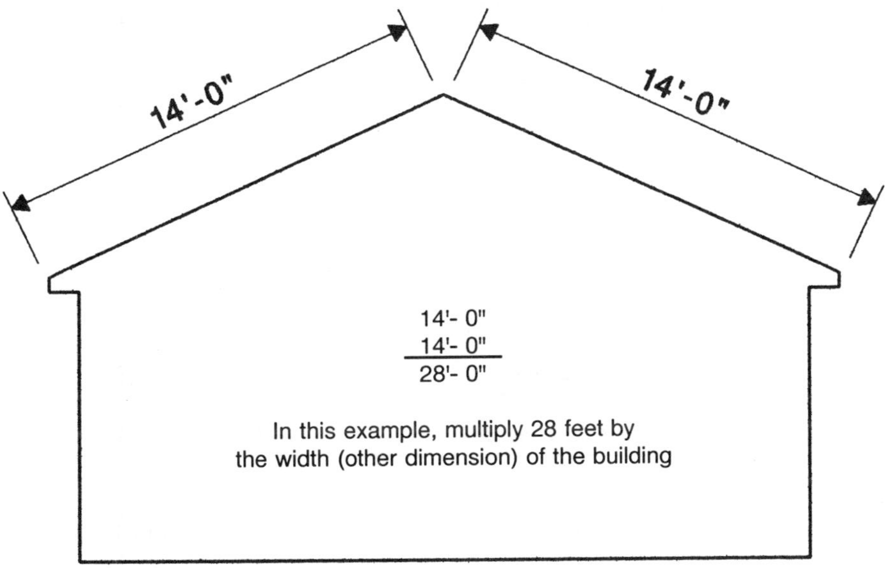

9. ASPHALT OR FIBERGLASS SHINGLES

SPECIFICATIONS			UNIT	PRICE	LOCAL AREA MODIFICATION %	LOCAL AREA MODIFICATION PRICE
	ALL ROOFING INSTALLATIONS ON THIS AND THE FOLLOWING PAGES ARE ON ONE STORY BUILDINGS					
MORE THAN 1 STORY ABOVE GRADE	FOR ROOFING MORE THAN ONE STORY ABOVE GRADE					
	2 STORIES ABOVE GRADE	**ADD**	EA	90.00		
	3 STORIES ABOVE GRADE	**ADD**	EA	135.00		
SMALL JOBS	SMALL JOB ROOFING LESS THAN 300 SF	**ADD**	EA	105.00		
	300 TO 500 SF	**ADD**	EA	60.00		
	500 TO 600 SF	**ADD**	EA	30.00		
DIFFICULT ROOF COVERING	DIFFICULT ROOF COVERING, CUT UP WITH VALLEYS AND DORMERS					
	SOME DIFFICULTY	**ADD**	SF	.30		
	DIFFICULT	**ADD**	SF	.60		
	VERY DIFFICULT	**ADD**	SF	.90		
ASPHALT OR FIBERGLASS SHINGLES, 215-225 LB (20 YR)	ON PLAIN SHED OR GABLE WOOD ROOF DECK • LOAD ONTO ROOF • BUILD STAGING IF REQUIRED • 3 TAB SQUARE BUTT SELF SEAL 215-225 LB. • 15 LB. FELT PAPER • GALVANIZED SHINGLE NAILS OR STAPLES • METAL DRIP EDGE 5"					
	Slope					
	4 TO 6 IN 12		SF	1.34		
	7 TO 12 IN 12		SF	1.53		
250 LB. (25 YR.)	***Slope***					
	4 TO 6 IN 12		SF	1.37		
	7 TO 12 IN 12		SF	1.56		
260 LB. (25 YR.)	***Slope***					
	4 TO 6 IN 12		SF	1.65		
	7 TO 12 IN 12		SF	1.85		
300 LB. (30 YR.)	***Slope***					
	4 TO 6 IN 12		SF	2.09		
	7 TO 12 IN 12		SF	2.28		
375 LB. (40 YR.)	***Slope***					
	4 TO 6 IN 12		SF	2.21		
	7 TO 12 IN 12		SF	2.40		

SPECIFICATIONS		UNIT	PRICE	LOCAL AREA MODIFICATION	
				%	PRICE
ROLL ROOFING	ON WOOD ROOF DECK WITH AVERAGE AMOUNT OF FITTING • ONE STORY ABOVE GRADE • MINERAL SURFACED 90 LB. • **NO** UNDERLAY • EXPOSED NAILS • TARRED SEAMS • 32" EXPOSURE (4" LAP) ***Slope***				
	4 TO 6 IN 12	SF	.62		
	7 TO 12 IN 12	SF	.65		
SELVAGE	ON WOOD ROOF DECK WITH AVERAGE AMOUNT OF FITTING • ONE STORY ABOVE GRADE • 19" SELVAGE, 110 TO 120 LB. • **NO** UNDERLAY • 17" EXPOSURE, 19" TO LAP • TARRED SEAMS ***Slope***				
	4 TO 6 IN 12	SF	.95		
	7 TO 12 IN 12	SF	.99		
OTHER TYPE ROOFS	INSTALL ASPHALT OR FIBERGLASS ROOFING ON OTHER TYPE ROOFS IN PLACE OF PLAIN SHED OR GABLE ROOFS				
	SMALL GABLE DORMER **ADD**	SF	50%		
	GAMBREL **ADD**	SF	25%		
	HIP **ADD**	SF	22%		
	NO FELT PAPER **DEDUCT**	SF	.12		
	NO METAL DRIP EDGE **DEDUCT**	SF	.12		
ICE DAM BARRIER	ICE AND WATER SHIELD 36" WIDE, RUBBERIZED MATERIAL PLACED UNDER STARTER COURSE AND IN VALLEYS	LF	2.97		

9. SINGLE PLY & BUILT-UP ROOFS ON WOOD DECK

SPECIFICATIONS		UNIT	PRICE	LOCAL AREA MODIFICATION	
				%	PRICE
	THE ROOFING PRICES SHOWN ON THIS PAGE INCLUDE A ROOFING SUBCONTRACTOR'S OVERHEAD AND PROFIT				
SINGLE PLY	"RUBBERIZED" OVER WOOD ROOF DECK • SINGLE PLY FLEXIBLE SHEET MEMBRANE • MINIMUM SLOPE 1/8" • MAXIMUM SLOPE 4"	EA PLUS SF	300.00 3.60		
BITUMEN	MODIFIED BITUMEN REINFORCED WITH FABRIC MESH, HEAT APPLICATION (TORCH APPLIED) • MINIMUM SLOPE 1/8" • MAXIMUM SLOPE 4"	EA PLUS SF	300.00 3.15		
2 PLY	• #30 FELT NAILED TO DECK • 2 PLY #15 FELT • 3 COATS HOT ASPHALT • GRAVEL, SLAG OR MARBLE CHIPS • MINIMUM SLOPE 1/8" • MAXIMUM SLOPE 4"	EA PLUS SF	472.50 2.93		
3 PLY	• #30 FELT NAILED TO DECK • 3 PLY #15 FELT • 4 MOP COATS HOT ASPHALT • GRAVEL, SLAG OR MARBLE CHIPS • MINIMUM SLOPE 1/8" • MAXIMUM SLOPE 4"	EA PLUS SF	487.50 3.15		
4 PLY	• #30 FELT NAILED TO DECK • 4 PLY #15 FELT • 5 MOP COATS HOT ASPHALT • GRAVEL, SLAG OR MARBLE CHIPS • MINIMUM SLOPE 1/8" • MAXIMUM SLOPE 4"	EA PLUS SF	487.50 3.60		

SPECIFICATIONS			UNIT	PRICE	LOCAL AREA MODIFICATION %	LOCAL AREA MODIFICATION PRICE
	SLATE ROOF PRICES BELOW ARE FOR ONE STORY BUILDINGS					
MORE THAN 1 STORY ABOVE GRADE	FOR ROOFING MORE THAN ONE STORY ABOVE GRADE					
	2 STORIES ABOVE GRADE	**ADD**	EA	90.00		
	3 STORIES ABOVE GRADE	**ADD**	EA	135.00		
SMALL JOBS	SMALL JOB ROOFING LESS THAN 300 SF	**ADD**	EA	105.00		
	300 TO 500 SF	**ADD**	EA	60.00		
	500 TO 600 SF	**ADD**	EA	30.00		
DIFFICULT ROOF COVERING	DIFFICULT ROOF COVERING, CUT UP WITH VALLEYS AND DORMERS					
	SOME DIFFICULTY	**ADD**	SF	.45		
	DIFFICULT	**ADD**	SF	.75		
	VERY DIFFICULT	**ADD**	SF	1.20		
BANGOR SLATE	ON PLAIN SHED OR GABLE WOOD ROOF DECK, 10" X 18" X 3/16" SLATE, 3" LAP, 7-1/2" EXPOSURE • LOAD ONTO ROOF • 1-1/4" COPPER SLATING NAILS • BUILD STAGING IF REQUIRED • 30 LB. ASPHALT FELT PAPER • RIDGE SLATES					
	Slope					
	4 TO 6 IN 12		SF	7.50		
	7 TO 12 IN 12		SF	7.97		
PENNSYLVANIA SLATE	SAME AS ABOVE, PENNSYLVANIA SLATE					
	Slope					
	4 TO 6 IN 12		SF	8.93		
	7 TO 12 IN 12		SF	9.39		
VERMONT SLATE	SAME AS ABOVE, VERMONT SLATE					
	Slope					
	4 TO 6 IN 12		SF	8.93		
	7 TO 12 IN 12		SF	9.39		
OTHER TYPE ROOFS	INSTALL SLATE ROOFING ON OTHER TYPE ROOFS IN PLACE OF PLAIN SHED OR GABLE ROOFS					
	SMALL GABLE DORMER	**ADD**	SF	50%		
	GAMBREL	**ADD**	SF	25%		
	HIP	**ADD**	SF	22%		

9. CEDAR SHAKE ROOFING

SPECIFICATIONS			UNIT	PRICE	LOCAL AREA MODIFICATION %	LOCAL AREA MODIFICATION PRICE
	CEDAR SHINGLE AND HANDSPLIT SHAKE PRICES ARE FOR ONE STORY BUILDINGS					
MORE THAN 1 STORY ABOVE GRADE	FOR ROOFING MORE THAN ONE STORY ABOVE GRADE					
	2 STORIES ABOVE GRADE	**ADD**	EA	90.00		
	3 STORIES ABOVE GRADE	**ADD**	EA	135.00		
SMALL JOBS	SMALL JOB ROOFING LESS THAN 300 SF	**ADD**	EA	105.00		
	300 TO 500 SF	**ADD**	EA	60.00		
	500 TO 600 SF	**ADD**	EA	30.00		
DIFFICULT ROOF COVERING	DIFFICULT ROOF COVERING, CUT UP WITH VALLEYS AND DORMERS					
	SOME DIFFICULTY	**ADD**	SF	.45		
	DIFFICULT	**ADD**	SF	.75		
	VERY DIFFICULT	**ADD**	SF	1.20		
HANDSPLIT AND RESAWN, 24"	ON PLAIN SHED OR GABLE TYPE WOOD ROOF DECK					
	HANDSPLIT AND RESAWN CEDAR SHAKES, 24" LONG, 10" EXPOSURE, 1/2" TO 3/4" THICKNESS • 15 LB. FELT PAPER BETWEEN COURSES • 6d RUST RESISTANT NAILS • STAGING AS REQUIRED					
	Slope					
	4 TO 6 IN 12		SF	4.67		
	7 TO 12 IN 12		SF	5.15		
	SAME AS ABOVE, 3/4" TO 1-1/4" THICKNESS					
	Slope					
	4 TO 6 IN 12		SF	5.13		
	7 TO 12 IN 12		SF	5.63		

SPECIFICATIONS		UNIT	PRICE	LOCAL AREA MODIFICATION %	LOCAL AREA MODIFICATION PRICE
16" CEDAR SHINGLES	ON PLAIN SHED OR GABLE TYPE WOOD ROOF DECK CEDAR SHINGLES 16" LONG, 5" EXPOSURE • 15 LB. FELT PAPER OVER SOLID WOOD ROOF SHEATHING OR **NO** FELT PAPER IF OVER SPACED SHEATHING • 3d RUST RESISTANT NAILS • STAGING AS REQUIRED ***Slope***				
	4 TO 6 IN 12	SF	5.18		
	7 TO 12 IN 12	SF	5.49		
18" CEDAR SHINGLES	SAME AS ABOVE, No. 1, 18" PERFECTIONS, 5-1/2" EXPOSURE ***Slope***				
	4 TO 6 IN 12	SF	5.27		
	7 TO 12 IN 12	SF	5.58		
OTHER TYPE ROOFS	INSTALL CEDAR SHINGLES AND HANDSPLIT SHAKES ON OTHER TYPE ROOFS IN PLACE OF PLAIN SHED OR GABLE ROOFS				
	SMALL GABLE DORMER **ADD**	SF	50%		
	GAMBREL **ADD**	SF	25%		
	HIP **ADD**	SF	22%		

9. WOOD FIBER ROOFING SHINGLE PANELS

SPECIFICATIONS			UNIT	PRICE	LOCAL AREA MODIFICATION %	LOCAL AREA MODIFICATION PRICE
	ROOFING SHINGLE AND SHAKE PANEL PRICES ARE FOR ONE STORY BUILDINGS					
MORE THAN 1 STORY ABOVE GRADE	FOR ROOFING MORE THAN ONE STORY ABOVE GRADE					
	2 STORIES ABOVE GRADE	**ADD**	EA	90.00		
	3 STORIES ABOVE GRADE	**ADD**	EA	135.00		
SMALL JOBS	SMALL JOB ROOFING LESS THAN 300 SF	**ADD**	EA	105.00		
	300 TO 500 SF	**ADD**	EA	60.00		
	500 TO 600 SF	**ADD**	EA	30.00		
DIFFICULT ROOF COVERING	DIFFICULT ROOF COVERING, CUT UP WITH VALLEYS AND DORMERS					
	SOME DIFFICULTY	**ADD**	SF	.30		
	DIFFICULT	**ADD**	SF	.60		
	VERY DIFFICULT	**ADD**	SF	.90		
MASONITE PANEL	• ON EXISTING PLAIN SHED OR GABLE WOOD ROOF DECK OR OVER ONE LAYER OF OLD ROOF • GALVANIZED ROOFING NAILS OR STAPLES • #30 FELT PAPER					
	Traditional L 48" X 11-25/32" 230 LBS/SQ 9-1/4" EXPOSURE ***Slope***					
	4 TO 6 IN 12		SF	2.18		
	7 TO 12 IN 12		SF	2.57		
	Traditional C 48" X 11-25/32" 265 LBS/SQ 9-1/4" EXPOSURE ***Slope***					
	4 TO 6 IN 12		SF	2.61		
	7 TO 12 IN 12		SF	3.00		
OTHER TYPE ROOFS	WOOD FIBER SHINGLE PANELS ON OTHER TYPE ROOFS IN PLACE OF PLAIN SHED OR GABLE ROOFS					
	SMALL GABLE DORMER	**ADD**	SF	50%		
	GAMBREL	**ADD**	SF	25%		
	HIP	**ADD**	SF	22%		

9. HANDSPLIT CEDAR SHAKE ROOF PANELS

SPECIFICATIONS		UNIT	PRICE	LOCAL AREA MODIFICATION	
				%	PRICE
HANDSPLIT ROOF PANELS	• ON EXISTING RAFTERS OR EXISTING PLAIN SHED OR GABLE ROOF DECK • 24" X 96" HANDSPLIT ROOF PANEL • No. 1 GRADE CEDAR SHAKES BONDED TO 5" X 1/2" PLYWOOD STRIPS • HIP AND RIDGE UNITS AS REQUIRED • #30 LB. FELT PAPER BETWEEN COURSES • 10" EXPOSURE • RUST RESISTANT NAILS				
	Slope				
	4 TO 6 IN 12	SF	3.69		
	7 TO 12 IN 12	SF	4.08		
OTHER TYPE ROOFS	INSTALL HANDSPLIT ROOF PANELS ON OTHER TYPE ROOFS IN PLACE OF PLAIN SHED OR GABLE ROOFS				
	SMALL GABLE DORMER **ADD**	SF	50%		
	GAMBREL **ADD**	SF	25%		
	HIP **ADD**	SF	22%		

9. CONCRETE ROOFING TILES

	SPECIFICATIONS		UNIT	PRICE	LOCAL AREA MODIFICATION %	LOCAL AREA MODIFICATION PRICE
	CONCRETE ROOFING TILE PRICES ARE FOR ONE STORY BUILDINGS					
MORE THAN 1 STORY ABOVE GRADE	FOR ROOFING MORE THAN ONE STORY ABOVE GRADE					
	2 STORIES ABOVE GRADE	**ADD**	EA	90.00		
	3 STORIES ABOVE GRADE	**ADD**	EA	135.00		
SMALL JOBS	SMALL JOB ROOFING LESS THAN 300 SF	**ADD**	EA	105.00		
	300 TO 500 SF	**ADD**	EA	60.00		
	500 TO 600 SF	**ADD**	EA	30.00		
DIFFICULT ROOF COVERING	DIFFICULT ROOF COVERING, CUT UP WITH VALLEYS AND DORMERS SOME DIFFICULTY	**ADD**	SF	.30		
	DIFFICULT	**ADD**	SF	.60		
	VERY DIFFICULT	**ADD**	SF	.90		
CONCRETE ROOFING TILES (FLAT)	ON PLAIN SHED OR GABLE SOLID SHEATHING WOOD ROOF DECK • 11" X 17" X 1-1/4" THICK ROOFING TILES WITH LUGS AT HEAD • 2" WOODEN LATH STRIPS NAILED OVER EACH RAFTER ITS FULL LENGTH • 1" X 2" REDWOOD BATTENS NAILED ACROSS LATH STRIPS SPACED TO HANG TILES • NAIL EVERY 3rd COURSE OF TILE ***Slope*** UNDER 3 IN 12		SF	4.14		
	OVER 3 IN 12 (OMIT HOT MOPPING)		SF	3.57		
	ON EXISTING 1" X 6" SPACED SHEATHING (SPACED TO HANG TILES) ONE STORY ABOVE GRADE • INTERLACE 16" STRIPS OF 30 LB. ROOFING FELT BETWEEN EACH COURSE ***Slope*** OVER 4 IN 12		SF	2.73		
OTHER TYPE ROOFS	CONCRETE ROOFING TILES ON OTHER TYPE ROOFS IN PLACE OF PLAIN SHED OR GABLE ROOFS SMALL GABLE DORMER	**ADD**	SF	50%		
	GAMBREL	**ADD**	SF	25%		
	HIP	**ADD**	SF	22%		

SPECIFICATIONS		UNIT	PRICE	LOCAL AREA MODIFICATION %	LOCAL AREA MODIFICATION PRICE
	MINERAL FIBER TYPE SHINGLE PRICES ARE FOR ONE STORY BUILDINGS				
MORE THAN 1 STORY ABOVE GRADE	FOR ROOFING MORE THAN ONE STORY ABOVE GRADE				
	2 STORIES ABOVE GRADE **ADD**	EA	90.00		
	3 STORIES ABOVE GRADE **ADD**	EA	135.00		
SMALL JOBS	SMALL JOB ROOFING LESS THAN 300 SF **ADD**	EA	105.00		
	300 TO 500 SF **ADD**	EA	60.00		
	500 TO 600 SF **ADD**	EA	30.00		
DIFFICULT ROOF COVERING	DIFFICULT ROOF COVERING, CUT UP WITH VALLEYS AND DORMERS SOME DIFFICULTY **ADD**	SF	.30		
	DIFFICULT **ADD**	SF	.60		
	VERY DIFFICULT **ADD**	SF	.90		
SUPRADUR	ON PLAIN SHED OR GABLE SOLID SHEATHING WOOD ROOF DECK • 9-7/16" X 16" X 1/4" THICK MINERAL FIBER ROOF SHINGLES • 500 LBS/SQUARE • 7" EXPOSURE • #15 FELT PAPER • RUST RESISTANT NAILS ***Slope*** 3 TO 6 IN 12	SF	6.93		
	7 TO 12 IN 12	SF	7.32		
	• 14" X 30" X 3/16" THICK MINERAL FIBER ROOF SHINGLES • 325 LBS/SQUARE • 6" EXPOSURE • #15 FELT PAPER • RUST RESISTANT NAILS ***Slope*** 4 TO 6 IN 12	SF	6.78		
	7 TO 12 IN 12	SF	7.17		
OTHER TYPE ROOFS	MINERAL FIBER SHINGLES ON OTHER TYPE ROOFS IN PLACE OF PLAIN SHED OR GABLE ROOFS	SF	50%		
	SMALL GABLE DORMER **ADD**				
	GAMBREL **ADD**	SF	25%		
	HIP **ADD**	SF	22%		

9

9. CLAY ROOFING TILES

SPECIFICATIONS		UNIT	PRICE	LOCAL AREA MODIFICATION %	LOCAL AREA MODIFICATION PRICE
SPANISH TILE	ON PLAIN SHED OR GABLE WOOD ROOF DECK • 11" O.C., 15" EXPOSURE, 3" LAP, RED CLAY TILE • 30 LB. FELT PAPER • HOT MOP SURFACE • RUST RESISTANT NAILS • INCLUDING RIDGE AND ACCESSORY TILES ***Slope***				
	3 TO 6 IN 12	SF	8.76		
	7 TO 12 IN 12	SF	9.35		
MISSION (BARREL) TILE	ON WOOD ROOF DECK, 18" TILE, 10-3/4" O.C., 14" EXPOSURE, 4" LAP • 30 LB. FELT PAPER • 1" X 4" WOOD NAILING STRIPS • HOT MOP SURFACE • RUST RESISTANT NAILS • INCLUDING RIDGE AND ACCESSORY TILES ***Slope***				
	3 TO 6 IN 12	SF	11.40		
	7 TO 12 IN 12	SF	11.99		
	WOOD BATTEN STRIP UNDER	SF	.51		
OTHER TYPE ROOFS	CLAY ROOFING TILES ON OTHER TYPE ROOFS IN PLACE OF PLAIN SHED OR GABLE ROOFS				
	SMALL GABLE DORMER **ADD**	SF	50%		
	GAMBREL **ADD**	SF	25%		
	HIP **ADD**	SF	22%		

SPECIFICATIONS		UNIT	PRICE	LOCAL AREA MODIFICATION %	LOCAL AREA MODIFICATION PRICE
	ROOFING PRICES SHOWN ON THIS PAGE INCLUDE A ROOFING SUBCONTRACTOR'S OVERHEAD AND PROFIT				
COPPER	OVER WOOD DECK • 16 OZ. (.020 INCH) COPPER • #15 FELT PAPER				
	STANDING SEAM	EA PLUS SF	615.00 15.06		
	FLAT SEAM	EA PLUS SF	615.00 13.95		
	EA = EACH JOB LOCATION				
METAL	OVER WOOD DECK • 30 GA. SHEET METAL WITH BAKED ENAMEL FINISH • #15 FELT PAPER				
	STANDING SEAM	EA PLUS SF	360.00 4.65		
	FLAT SEAM	EA PLUS SF	360.00 4.50		
	EA = EACH JOB LOCATION				
	NOTE: FOR ALUMINIZATION OF METAL ROOFS, SEE PAGE 146				

9. ROOFING TEAR-OFF

SPECIFICATIONS		UNIT	PRICE	LOCAL AREA MODIFICATION %	LOCAL AREA MODIFICATION PRICE
	FOR RE-ROOFING, USE THE SAME PRICES SHOWN FOR NEW ROOFS ON PAGES 132-143. PRICES FOR TEARING OFF THE OLD ROOF, DEPOSITING DEBRIS DIRECTLY INTO TRUCK FROM ROOF AND HAULING TO DUMP WITHIN 5 MILES (**DUMPING FEE NOT INCLUDED**) ARE SHOWN BELOW FOR VARIOUS KINDS OF EXISTING ROOFS.				
ASPHALT OR FIBERGLASS	ASPHALT OR FIBERGLASS SHINGLES — 1 LAYER				
	UP TO 8 IN 12 SLOPE	SF	.41		
	STEEP SLOPE	SF	.53		
	ASPHALT OR FIBERGLASS SHINGLES — 2 LAYERS				
	UP TO 8 IN 12 SLOPE	SF	.59		
	STEEP SLOPE	SF	.69		
	ASPHALT OR FIBERGLASS SHINGLES — 3 LAYERS				
	UP TO 8 IN 12 SLOPE	SF	.75		
	STEEP SLOPE	SF	.87		
ROLL ROOF OR SELVAGE	ROLL ROOFING OR SELVAGE				
	UP TO 8 IN 12 SLOPE	SF	.33		
	STEEP SLOPE	SF	.41		
BUILT-UP	3, 4 OR 5 PLY ROOF				
	WITH GRAVEL	SF	1.34		
	WITHOUT GRAVEL	SF	.69		
SLATE	UP TO 8 IN 12 SLOPE	SF	.83		
	STEEP SLOPE	SF	.92		
CLAY TILE	1-PIECE OR 2-PIECE INTERLOCKING				
	UP TO 8 IN 12 SLOPE	SF	.69		
	STEEP SLOPE	SF	.84		
METAL ROOFING	COPPER, GALVANIZED OR ALUMINUM ROOFING	SF	.47		
CEDAR SHINGLES	16" OR 18" SHINGLES WITH 5" EXPOSURE				
	UP TO 8 IN 12 SLOPE	SF	.56		
	STEEP SLOPE	SF	.69		
CEDAR SHAKES	24" CEDAR SHAKES WITH 10" EXPOSURE, 3/4" TO 5/4"				
	UP TO 8 IN 12 SLOPE	SF	.38		
	STEEP SLOPE	SF	.53		

9

SPECIFICATIONS		UNIT	PRICE	LOCAL AREA MODIFICATION	
				%	PRICE
PREPARATION FOR RE-ROOFING	INSTALL 1/2" PLYWOOD OVER EXISTING ROOF COVERING BEFORE INSTALLING NEW ROOF	SF	1.26		
	CUT BACK EXISTING ROOFING FROM EAVES AND/OR GABLE, AND FILL SPACE WITH 1" X 6" FIR OR PINE LF = EAVES AND/OR GABLE	LF	1.23		
	NAIL BATTENS OVER EXISTING ROOF COVERING TO RECEIVE NEW ROOF COVERING				
	1" X 3"	LF	.45		
	1" X 4"	LF	.53		
	1" X 6"	LF	.63		
	LF = BATTENS				

9. ROOF REPAIR

SPECIFICATIONS		UNIT	PRICE	LOCAL AREA MODIFICATION	
				%	PRICE
PATCH BUILT-UP ROOF	SCRAPE MATERIALS FROM DEFECTIVE AREA AND INSTALL NEW PAPER, HOT ASPHALT AND SLAG TO AREA	EA PLUS SF	135.00 1.67		
METAL ROOF REPAIR	TAR OVER EXISTING METAL ROOF FOR TEMPORARY REPAIR OF LEAKS	SF	1.68		
ALUMINIZATION	• SCRAPE MATERIALS FROM DEFECTIVE AREA • PATCH HOLES WITH FIBER TAPE AND BLACK ASPHALT ROOF CEMENT • APPLY 2 COATS OF ALUMINUM ROOF COATING WITH FIBER (2nd COAT WHILE 1st COAT IS STILL TACKY)	SF	6.95		
	SAME AS ABOVE, DIFFICULT CONDITIONS — STEEP ROOF SLOPE, ETC.	SF	9.62		
	ALUMINUM BASED ROOF PAINT APPLIED ON NEW ROOF OR ROOF IN GOOD CONDITION, ONE COAT EACH 2 YEARS SF = ONE COAT	SF	1.95		
	SAME AS ABOVE, DIFFICULT CONDITIONS — STEEP ROOF SLOPE, ETC.	SF	2.36		
SLATE ROOF REPAIR	REPAIR SLATE ROOF LEAK: REMOVE INDIVIDUAL SLATE AND REPLACE WITH NEW EA = UP TO 10 SLATES	EA	291.00		
	FOR EACH SLATE MORE THAN 10 **ADD**	EA	15.00		
SLATE	REPLACE VENT PIPE COLLAR WITH NEW LEAD COLLAR	EA	243.00		
SLATE VALLEY	REPAIR SLATE VALLEY ON DORMER EA = BOTH VALLEYS	EA PLUS LF	483.00 81.90		
	REPAIR JOB: CEMENT RIDGE, REPLACE 5 SLATES, SEAL AROUND CHIMNEY AND VENTS EA = ENTIRE JOB	EA	411.00		
ASBESTOS ROOFING	REMOVE DEFECTIVE ASBESTOS SHINGLES AND REPLACE WITH NEW (IF AVAILABLE) EA = UP TO 10 SHINGLES	EA	237.00		
CLAY TILE ROOF REPAIR	REMOVE DEFECTIVE FLAT OR CURVED TILES AND REPLACE WITH NEW TILES EA = UP TO 10 TILES	EA	237.00		

SPECIFICATIONS		UNIT	PRICE	LOCAL AREA MODIFICATION %	LOCAL AREA MODIFICATION PRICE
ASPHALT OR FIBERGLASS SHINGLE REPAIR	REMOVE DEFECTIVE ASPHALT OR FIBERGLASS SHINGLES AND REPLACE WITH NEW EA = UP TO 10 SHINGLES	EA	153.00		
	FOR EACH SHINGLE MORE THAN 10 **ADD**	EA	8.55		
REPAIR WOOD SHINGLES	REPAIR WOOD SHINGLE OR SHAKE LEAK BY NAIL-ING DOWN, INSTALLING SHEET METAL PATCH UNDER OR CEMENTING CRACK EA = UP TO 10 SHINGLES OR SHAKES	EA	99.00		
	FOR EACH SHINGLE OR SHAKE MORE THAN 10 **ADD**	EA	6.45		
	REPAIR SHINGLE OR SHAKE LEAK BY REPLACING DEFECTIVE PIECES WITH NEW ONES EA = UP TO 10 SHINGLES OR SHAKES	EA	156.00		
	FOR EACH SHINGLE OR SHAKE MORE THAN 10 **ADD**	EA	8.10		

9. FLASHING AND ROOF VENTS

SPECIFICATIONS		UNIT	PRICE	LOCAL AREA MODIFICATION	
				%	PRICE
FLASH CHIMNEY	FLASH AVERAGE SIZE EXTERIOR OR INTERIOR CHIMNEY WITH STEPFLASHING AND COUNTERFLASHING TURNED 1/2" INTO OPEN JOINTS, **NOT** INCLUDING CUTTING JOINTS				
	ALUMINUM, .032 GA.	EA	75.00		
	COPPER, 16-20 OZ.	EA	138.00		
FLASH ROOF	FLASH GABLE ROOF TO WALL WITH STEP-FLASHING				
	In Brick Open Joints				
	ALUMINUM	LF	7.73		
	COPPER	LF	15.90		
	On Wood Sheathing				
	ALUMINUM	LF	9.00		
	COPPER	LF	16.50		
	FLASH TOP OF FLAT OR SHED ROOF TO WALL IN OPEN JOINTS OR ON WOOD SHEATHING				
	ALUMINUM	LF	5.10		
	COPPER	LF	9.83		
	VALLEY FLASHING, 18" WIDE				
	ALUMINUM	LF	10.20		
	COPPER	LF	15.30		
CUT JOINTS	CUT JOINTS IN MASONRY FOR FLASHING				
	SOFT MORTAR	LF	2.25		
	HARD MORTAR	LF	3.90		
PARAPET	RE-FLASH PARAPET 12" HIGH AND 12" WIDE				
	ALUMINUM	LF	8.70		
	COPPER	LF	24.00		
PLUMBING FLASHING	REPLACE EXISTING FLASHING ON PLUMBING VENT	EA	34.80		
ROOF VENTILATOR	TURBINE VENTILATOR WITH WIND-DIRECTED ROTATING TOP, 12"	EA	94.50		
ROOF VENT	UP TO 9" DIAMETER SCREENED PLASTIC ROOF VENT IN FIBERGLASS, ROLL ROOFING & WOOD SHINGLE ROOFS	EA	49.50		
RIDGE VENT	CONTINUOUS ROOF RIDGE VENT WITH LOUVERED SIDE OPENINGS -- INCLUDES CUTTING EXISTING SHINGLES AND SHEATHING				
	ALUMINUM PAINTED	LF	9.08		
	VINYL COLORED	LF	9.23		
	SAME AS ABOVE, INSTALLATION ONLY				
	ALUMINUM PAINTED	LF	7.20		
	VINYL COLORED	LF	7.35		

9

9. GUTTERS AND DOWNSPOUTS

SPECIFICATIONS		UNIT	PRICE	LOCAL AREA MODIFICATION	
				%	PRICE
ALUMINUM GUTTERS AND DOWN-SPOUTS	ALUMINUM GUTTER AND DOWNSPOUT SYSTEM WITH BAKED-ON WHITE ENAMEL FINISH, INCLUDING ACCESSORIES LF = TOTAL COMBINED GUTTER AND DOWN-SPOUT LENGTH	LF	4.68		
	.032 SEAMLESS GUTTER AND DOWNSPOUT, WHITE ENAMEL FINISH	LF	5.96		
GALVANIZED GUTTERS & DOWN-SPOUTS	GALVANIZED STEEL OG GUTTER AND DOWN-SPOUT, INCLUDING ACCESSORIES				
	5"	LF	4.14		
	6"	LF	5.21		
	LF = TOTAL COMBINED GUTTER AND DOWN-SPOUT LENGTH				
COPPER	HALF ROUND COPPER GUTTER AND DOWN-SPOUTS, INCLUDING ACCESSORIES				
	5"	LF	13.71		
	6"	LF	16.65		
	LF = TOTAL COMBINED GUTTER AND DOWN-SPOUT LENGTH				
WOOD	4" X 5" TREATED FIR OR HEMLOCK	LF	18.54		
VINYL	VINYL 5" COLORED OR WHITE GUTTER AND DOWNSPOUT, INCLUDING ACCESSORIES LF = TOTAL COMBINED GUTTER AND DOWN-SPOUT LENGTH	LF	4.07		
GUTTER GUARD	ALUMINUM MESH GUTTER GUARD, 6" WIDE	LF	1.71		
CANALES	CANALES, FOR FLAT ROOF APPLICATION ONLY	EA	63.00		

9

10. EXTERIOR TRIM

SPECIFICATIONS		UNIT	PRICE	LOCAL AREA MODIFICATION	
				%	PRICE
FASCIA OR FRIEZE	#2 PINE, SPRUCE, HEMLOCK, FIR				
	1 X 6	LF	2.70		
	1 X 8	LF	3.14		
	1 X 10	LF	3.62		
	1 X 12	LF	4.31		
SOFFIT	#2 PINE, SPRUCE, HEMLOCK, FIR				
	6"	LF	3.14		
	8"	LF	3.62		
	10"	LF	4.01		
	12"	LF	4.76		
	3/8" FIR PLYWOOD, AC EXTERIOR GRADE				
	12"	LF	3.56		
	16"	LF	4.19		
	18"	LF	5.03		
	24"	LF	5.51		
	30"	LF	7.85		
	36"	LF	8.33		
	48"	LF	7.70		
BED MOULD	PINE, FIR, HEMLOCK, 3/4" X 2-5/8"	LF	2.10		
RAKE	#2 PINE OR FIR RAKE 1 X 6 AND RAKE MOULD (SHINGLE MOULD)	LF	3.30		
ATTIC LOUVER	REDWOOD, INCLUDING INSECT SCREEN				
	CIRCULAR 24" DIAMETER	EA	168.00		
	HALF CIRCLE 30" X 18"	EA	174.00		
	RECTANGULAR 16" X 24"	EA	117.00		
LATTICE WALL	CUSTOM MADE WITH PRESSURE TREATED LUMBER, KILN DRIED				
	1/4" X 1-1/8", 2-1/4" O.C.	SF	16.49		
	1/4" X 1-5/8", 3-1/4" O.C.	SF	14.66		
LATTICE PANEL	LATTICE PANEL, 4' X 8', PRE-ASSEMBLED, WITH SLATS 1/4" X 1-1/2", PRESSURE TREATED	SF	2.84		
	PLASTIC PRE-FINISHED 4' X 8' LATTICE PANEL	SF	3.23		

SPECIFICATIONS		UNIT	PRICE	LOCAL AREA MODIFICATION %	LOCAL AREA MODIFICATION PRICE
ALUMINUM SOFFIT	RIBBED SOFFIT, INCLUDING "F" CHANNEL				
	8"	LF	5.10		
	12"	LF	5.85		
	18"	LF	7.20		
	24"	LF	8.48		
ALUMINUM FASCIA OR RAKE	FASCIA OR RAKE TRIM				
	4"	LF	3.60		
	6"	LF	3.83		
	8"	LF	4.19		
	10"	LF	4.50		
SOFFIT VENT	PLUG TYPE, ROUND, ALUMINUM, INSTALLED IN EXISTING SOFFIT				
	1"	EA	8.79		
	3"	EA	10.97		
	CONTINUOUS SOFFIT VENT, INSTALLED IN NEW SOFFIT, 3"	LF	3.98		
GRILLE	ALUMINUM GRILLE INSTALLED IN EXISTING SOFFIT, 8" X 12"	EA	36.00		
GABLE VENTS	GABLE LOUVER VENTS				
	CENTER HEIGHT: 12"	EA	99.00		
	16"	EA	111.00		
	20"	EA	126.00		
ATTIC LOUVER VENTS	INCLUDING INSECT SCREEN, INSTALLED IN EXISTING HOLE				
	16" X 4"	EA	27.00		
	16" X 8"	EA	36.00		
	15" X 21"	EA	51.00		
	24" X 30"	EA	66.00		
	33" X 27"	EA	78.00		
VINYL SOFFIT	SOLID OR PERFORATED, INCLUDING "F" OR "J" CHANNEL				
	12"	LF	4.73		
	18"	LF	5.93		
	24"	LF	6.98		
WRAP TRIM	WRAP EXTERIOR WINDOW OR DOOR TRIM WITH ALUMINUM EA = PER OPENING	EA	45.83		

10. WOOD DECK

SPECIFICATIONS		UNIT	PRICE	LOCAL AREA MODIFICATION	
				%	PRICE
	NOTE: RUST-RESISTANT NAILS ARE INCLUDED IN ALL DECK INSTALLATIONS				
WOOD POSTS	SOLID WOOD POSTS ON EXISTING FOOTING UP TO 8 FT.				
	4 X 4 PRESSURE TREATED FIR OR PINE	EA	31.41		
	4 X 4 CONSTR. GRADE REDWOOD OR TIGHT KNOT CEDAR	EA	40.86		
PIPE COLUMN	PIPE COLUMN, INCL. PLATES, 3" GALVANIZED, UP TO 8 FT.	EA	165.00		
WOOD BEAM OR HEADER	SOLID OR BUILT-UP ON EXISTING SUPPORTS, WITH PRESSURE TREATED LUMBER				
	4" X 8"	LF	5.55		
	6" X 8"	LF	7.77		
	4" X 10"	LF	6.90		
	6" X 10"	LF	10.32		
	4" X 12"	LF	8.40		
JOISTS	INCLUDING LEDGER BOLTED TO BUILDING, JOISTS DOUBLED AT ALL SIDES, JOIST HANGERS, WITH PRESSURE TREATED LUMBER, 16" O.C.				
	2" X 8"	SF	3.42		
	2" X 10"	SF	3.98		
	2" X 12"	SF	4.49		
BAND	CEDAR OR REDWOOD, 1" X 12"	LF	6.00		
DECK SURFACE	DECK APPLIED FLAT WITH 1/4" SPACE BETWEEN EACH PIECE				
	2" X 6" PRESSURE TREATED FIR OR PINE	SF	3.48		
	2" X 6" CONSTR. GRADE REDWOOD OR TIGHT KNOT CEDAR	SF	5.43		
	5/4" X 6" CONSTR. GRADE REDWOOD OR TIGHT KNOT CEDAR	SF	4.01		
	SAME AS ABOVE, DECK LAID DIAGONALLY				
	2" X 6" PRESSURE TREATED FIR OR PINE	SF	4.43		
	2" X 6" CONSTR. GRADE REDWOOD OR TIGHT KNOT CEDAR	SF	6.90		
	5/4" X 6" CONSTR. GRADE REDWOOD OR TIGHT KNOT CEDAR	SF	5.09		

SPECIFICATIONS		UNIT	PRICE	LOCAL AREA MODIFICATION %	LOCAL AREA MODIFICATION PRICE
DECK RAILING OR STEP RAILING	• 4" X 4" END AND INTERMEDIATE POSTS EXTENDING TO 42" ABOVE SURFACE OF DECK, 48" O.C. • 2" X 6" CAP LAID FLAT ON POSTS, 2" X 6" TOP RAIL UNDER CAP • 2" X 6" MIDDLE AND BOTTOM RAILS				
	PRESSURE TREATED FIR OR PINE	LF	12.30		
	CONSTR. GRADE REDWOOD OR CEDAR	LF	16.76		
	• 4" X 4" END AND INTERMEDIATE POSTS EXTENDING TO 42" ABOVE SURFACE OF DECK, 48" O.C. • 2" X 6" CAP LAID FLAT ON POSTS, 2" X 6" TOP RAIL UNDER CAP • 2" X 2" PICKETS 7-1/2" O.C. NAILED TO TOP RAIL & BAND				
	PRESSURE TREATED FIR OR PINE	LF	12.60		
	CONSTR. GRADE REDWOOD OR CEDAR	LF	17.18		
SEAT	ADD TO ABOVE RAILINGS, 18" SEAT 18" ABOVE DECK				
	PRESSURE TREATED FIR OR PINE	LF	25.38		
	CONSTR. GRADE REDWOOD OR CEDAR	LF	36.75		
STEPS TO DECK	2" X 12" STRINGERS TWO 2" X 6" PER TREAD, OPEN RISERS, CONCRETE BOTTOM TREAD ON GROUND				
	PRESSURE TREATED FIR OR PINE, 36" WIDE STEPS	PER STEP	40.13		
	CONSTR. GRADE REDWOOD OR CEDAR, 36" WIDE STEPS	PER STEP	64.50		
	PRESSURE TREATED FIR OR PINE, 48" WIDE STEPS	PER STEP	57.75		
	CONSTR. GRADE REDWOOD OR CEDAR, 48" WIDE STEPS	PER STEP	92.25		
	NOTE: STEP RAILING IS **NOT** INCLUDED IN COST				

10. PORCH

SPECIFICATIONS		UNIT	PRICE	LOCAL AREA MODIFICATION	
				%	PRICE
	NOTE: FOR PORCH FRAMING AND FINISHING ITEMS NOT SHOWN ON THIS PAGE, REFER TO APPROPRIATE SECTIONS OF THE MANUAL				
POSTS AND COLUMNS	BUILT-UP POST, 2 X 4 COVERED WITH FOUR 1 X 6, UP TO 8 FEET HIGH				
	#2 PINE OR FIR	EA	72.00		
	CONSTR. GRADE REDWOOD	EA	87.00		
	ALUMINUM ROUND COLONIAL PORTICO COLUMN				
	8"	LF	39.00		
	10"	LF	48.00		
	12"	LF	81.00		
	FIR HOLLOW ROUND COLONIAL PORTICO COLUMN				
	8"	LF	78.00		
	10"	LF	84.00		
	12"	LF	96.00		
PORCH CEILING	INCLUDES CEILING COVE AT WALL INTERSECTIONS				
	1" X 6" REDWOOD V-JOINT	SF	8.81		
	1" X 6" KNOTTY PINE V-JOINT	SF	4.97		
	3/8" FIR PLYWOOD CEILING INCLUDING PANEL STRIPS AND CEILING COVE	SF	1.95		
	FIR BEADED CEILING, CENTER BEADED ONE SIDE AND CENTER V-JOINT OTHER SIDE				
	5/8" X 4"	SF	3.48		
PORCH FLOOR	3/4" T&G PINE PORCH FLOORING	SF	6.26		
HEADER TRIM	#2 PINE OR FIR, 6" SOFFIT AND TWO 12" FASCIAS	LF	10.07		
SCREEN	ALUMINUM SCREENING WITH 2 X 4 INTERMEDIATE FRAMING AND PANEL STRIPS	SF	4.14		
SCREEN DOOR, WOOD	WOOD FRAME SCREEN DOOR WITH ALUMINUM SCREENING, INCLUDING ALL NECESSARY HARDWARE FOR INSTALLATION	EA	162.00		

SPECIFICATIONS		UNIT	PRICE	LOCAL AREA MODIFICATION %	LOCAL AREA MODIFICATION PRICE
ORNAMENTAL IRON	ORNAMENTAL IRON PORCH AND STEP RAILING, INCLUDING ALL POSTS & FITTINGS REQUIRED FOR INSTALLATION				
	PREMIUM	LF	33.83		
	ECONOMY	LF	29.93		
	CUSTOM BUILT AND INSTALLED IRON RAILING	LF	39.23		
	ORNAMENTAL IRON COLUMN, 1" SQUARE TUBING, UP TO 8 FT. HIGH				
	8"	EA	102.00		
	8" X 8"	EA	144.00		
CORRUGATED FIBERGLASS ROOF PANELS	NAILED TO WOOD FRAMING WITH RING SHANK NAILS WITH RUBBER WASHERS, 2-1/2" CORRUGATIONS, 8 FT., 10 FT., AND 12 FT. PANELS X 26"	SF	3.35		

11. WOOD SIDING

	SPECIFICATIONS	UNIT	PRICE	LOCAL AREA MODIFICATION %	LOCAL AREA MODIFICATION PRICE
	NOTE: ALL SIDING ITEMS ON THIS AND THE FOLLOWING PAGES INCLUDE SIDING, NAILS, PAPER, STARTER STRIPS AND ALL OTHER MATERIALS AND LABOR REQUIRED TO INSTALL ON EXISTING SHEATHED OR BRACED STUDWALL AND ON A ONE STORY BUILDING				
MORE THAN ONE STORY ABOVE GRADE	FOR INSTALLATIONS MORE THAN ONE STORY ABOVE GRADE				
	2 STORIES ABOVE GRADE **ADD**	SF	10%		
	3 STORIES ABOVE GRADE **ADD**	SF	20%		
HOUSE WRAP	SUBSTITUTE HOUSE WRAP FOR PAPER SPECIFIED IN SIDING INSTALLATIONS ON THIS AND THE FOLLOWING PAGES **ADD**	SF	.21		
BEVELED SIDING	• REDWOOD OR CEDAR BEVELED SIDING • #15 FELT OR ROSIN COATED PAPER • 5/4 X 4 OUTSIDE CORNERS • 5/4 X 5/4 INSIDE CORNERS				
	Cedar				
	1/2 X 4	SF	5.07		
	1/2 X 6	SF	4.41		
	1/2 X 8	SF	4.08		
	3/4 X 8	SF	4.43		
	3/4 X 10	SF	4.10		
	Redwood				
	1/2 X 4	SF	6.47		
	1/2 X 6	SF	5.73		
	1/2 X 8	SF	5.40		
	3/4 X 8	SF	5.88		
	3/4 X 10	SF	5.15		
GERMAN OR DROP SIDING	• DROP SIDING • #15 FELT OR ROSIN COATED PAPER • 3/4 X 4 OUTSIDE CORNERS • 3/4 X 3/4 INSIDE CORNERS				
	Fir/Pine				
	1 X 6	SF	3.23		
	1 X 8	SF	3.08		
	Cedar				
	1 X 6	SF	4.32		
	1 X 8	SF	4.19		
	Redwood				
	1 X 6	SF	5.52		
	1 X 8	SF	5.13		

SPECIFICATIONS		UNIT	PRICE	LOCAL AREA MODIFICATION %	LOCAL AREA MODIFICATION PRICE
TONGUE AND GROOVE	• #15 FELT OR ROSIN COATED PAPER • 3/4 X 4 OUTSIDE CORNERS				
	Cedar				
	1 X 4	SF	7.07		
	1 X 6	SF	6.51		
	1 X 8	SF	6.20		
	1 X 10	SF	5.72		
	Redwood				
	1 X 4	SF	9.45		
	1 X 6	SF	8.79		
	1 X 8	SF	8.42		
	1 X 10	SF	7.83		
	Fir/Pine				
	1 X 4	SF	4.47		
	1 X 6	SF	4.01		
	1 X 8	SF	3.72		
	1 X 10	SF	3.35		
BOARD AND BATTEN	• #15 FELT OR ROSIN COATED PAPER • NAILED TO EXISTING WALL SHEATHING OR 1" X 3" HORIZONTAL FURRING, 12" O.C. • 1" X 3" BATTENS				
	Cedar				
	1 X 8	SF	5.01		
	1 X 10	SF	4.76		
	1 X 12	SF	4.50		
	Redwood				
	1 X 8	SF	6.83		
	1 X 10	SF	6.48		
	1 X 12	SF	6.12		
DRIP CAP & WATER TABLE	DRIP CAP AND 1 X 4 WATER TABLE INSTALLED UNDER FIRST COURSE OF SIDING **ADD**	LF	1.19		
	LF = TOTAL LINEAL FEET OF COMBINATION DRIP CAP AND WATER TABLE				

11

11. CEDAR SHAKE AND SHINGLE SIDING

SPECIFICATIONS		UNIT	PRICE	LOCAL AREA MODIFICATION %	LOCAL AREA MODIFICATION PRICE
16" CEDAR SHINGLES	• 16" CEDAR SHINGLES • #15 FELT OR ROSIN COATED PAPER • 3d RUST RESISTANT NAILS • WOVEN OUTSIDE CORNERS • WOVEN INSIDE CORNERS				
	5" EXPOSURE	SF	4.31		
	7-1/2" EXPOSURE	SF	3.74		
	SAME AS ABOVE, FIRE RETARDANT SHINGLES				
	5" EXPOSURE	SF	5.28		
	7-1/2 EXPOSURE	SF	4.53		
18" PERFECTIONS	SAME AS ABOVE, NO. 1 18" PERFECTIONS				
	5-1/2" EXPOSURE	SF	4.95		
	7-1/2" EXPOSURE	SF	3.87		
	8-1/2" EXPOSURE	SF	3.50		
	SAME AS ABOVE, FIRE RETARDANT SHINGLES				
	5-1/2" EXPOSURE	SF	6.15		
	7-1/2" EXPOSURE	SF	4.82		
	8-1/2" EXPOSURE	SF	4.29		
	SAME AS ABOVE, DOUBLE COURSING • NO. 1 18" PERFECTIONS • UNDERCOURSE NO. 3 UTILITY GRADE • BUTT NAILED				
	14" EXPOSURE	SF	3.45		
HAND-SPLIT SHAKES	SAME AS ABOVE, 24" LONG HAND-SPLIT SHAKES				
	8-1/2" EXPOSURE	SF	4.01		
	10" EXPOSURE	SF	3.50		
	11-1/2" EXPOSURE	SF	3.02		
	SAME AS ABOVE, FIRE RETARDANT HAND-SPLIT SHAKES				
	8-1/2" EXPOSURE	SF	4.91		
	10" EXPOSURE	SF	4.25		
	11-1/2" EXPOSURE	SF	3.62		

SPECIFICATIONS		UNIT	PRICE	LOCAL AREA MODIFICATION	
				%	PRICE
PANEL SIDING	• NO. 1 RED CEDAR SHAKES OR SHINGLES BONDED TO UNDERCOURSE AND PLYWOOD • MATCHING CORNERS • MATCHING NAILS • #15 FELT PAPER				
	18" X 96" 14" EXPOSURE	SF	3.69		
	9" X 96" 7" EXPOSURE	SF	3.86		
CEDAR SHINGLE PANELS	• 16' LENGTHS • 7/16" THICKNESS • #15 FELT PAPER • 5/4" X 4" OUTSIDE CORNERS • 5/4" X 5/4" INSIDE CORNERS • CEDAR SHINGLE FINISH • PRIMED				
	12" WIDTH	SF	3.23		

11

11. HARDBOARD SIDING

SPECIFICATIONS		UNIT	PRICE	LOCAL AREA MODIFICATION	
				%	PRICE
LAP SIDING	• 16' LENGTHS • 7/16" THICKNESS • #15 FELT OR ROSIN COATED PAPER • 5/4" X 4" OUTSIDE CORNERS • 5/4" X 5/4" INSIDE CORNERS • ROUGH SAWN CEDAR FINISH • PRIMED				
	8" WIDTH	SF	2.48		
	12" WIDTH	SF	2.36		
	• 16' LENGTHS • 7/16" THICKNESS • #15 FELT PAPER • 5/4" X 4" OUTSIDE CORNERS • 5/4" X 5/4" INSIDE CORNERS • WOODGRAIN OR SMOOTH FINISHES • PREFINISHED				
	8" WIDTH	SF	2.76		
	12" WIDTH	SF	2.67		

Local Area Modificati
HomeTech Remodeling and Ren

Page	Item	Mod.
3	ADDITION	106*
5	ADDITION	106*
7	ADDITION	106*
9	BRICK	106
	BLOCK	105*
	PIERS	105*
11	ADDITION	104*
13	2nd STORY ADDN	102*
14-16	EXTRAS, ALLOW	99*
17	INTER. FINISH	100
18	EXTRAS, ALLOW	98
19	EXTRAS, ALLOW	97
21	REMODEL BATH	100
	NEW BATH	100
23	REMODEL BATH	100
	NEW BATH	100
25	REMODEL BATH	100
	NEW BATH	100
27	REMODEL BATH	101
	NEW BATH	100
28	AREA	100
	SUBSTITUTE	100
29	ELECTRICAL	99
	VINYL	97
30	BREEZEWAY	108*
31	CARPORT	110*
32	DECK	103*
33	PRESS. TR.	97*
	CEDAR/RW	113*
35	NO. 1	102
	NO. 2	103*
	NO. 3	103*
	NO. 4	103
	NO. 5	106*
	NO. 6	103
36	EXTRAS, ALLOW	98
37	EXTRAS, ALLOW	97*
38	INTER. FINISH	101
39	EXTRAS, ALLOW	98
40	GARAGE	106*
41	EXTRAS, ALLOW	98
43	REMODEL KITCHEN	99
	NEW KITCHEN	99
45	REMODEL KITCHEN	99
	NEW KITCHEN	99
47	REMODEL KITCHEN	99
	NEW KITCHEN	99
48	EXTRAS, ALLOW	99
49	COUNTER	99
	BULKHEAD	97
	PLUMBING	101
	ELECTRIC	99
	FLOOR	100
50	PORCH	106*
51	EXTRAS, ALLOW	104
52	REC ROOM	104
53	PARTITION	92*
	EXTRAS, ALLOW	97
54	CONVERT	105
	WINDOW	100
	VINYL	100
	FOOTING	102*
55	CONVERT	100
	EXTRAS, ALLOW	98
57	APARTMENT	103
	ALLOW	108*
59	APARTMENTS	104
	ALLOW	108*

Page	Item	Mod.
61	RENOVATE	102
	ALLOW	108*
62	REMOVE	104*
	RAILING	99
	FLOOR	104*
	REPAIR	97
	MUDJACK	103*
63	AREAWAY	105*
	CLEAR	95
	WEEP	96
	DRAIN	97
64	FIREWALL	94*
	SOUND	109*
	LEAD PAN	97
65	CLOSE IN	103*
	HEARTH	97*
	TRANSOM	107*
	PLASTER	104*
66	DOORS	97
	CASED	97
67	REMOVE	97*
	ADJUST	95
	HINGE	97
	LOCKSET	97
	ADJUST	106*
68	WINDOWS	98
	MASONRY	94
	WINDOWS	97
	MASONRY	94
69	GLAZE	97
	STUCK	95
	REPAIR	97
70	FIXED	97
	VENTING	97
	ACCESSOR.	99
	SHAFT	97
71	WINDOWS	97
72	COLUMN	120
	BEAM	100
	SILL	96
	DOUBLE	97*
	BEAM	98
74	PLANS	100
76	EQUIPMENT	90
	LABOR	95
77	LABOR	95
78	LABOR	95
79	STEEL	100
	TEAR-OUT	95
80 THROUGH 85		95
86	STEEL	95
	TEAR-OUT	95
87	TEAR-OUT	95
88	DISCONNECT	96
	SHOWER STALL	95
89	HEAT	96
	DUCT, A/C	95
90	TEAR-OUT	98
91	CLEAN/EXTERM.	95
92	DAMAGE	95
93	ABATEMENT	95
94	EXCAVATE	95
95	EXCAV., GRADE	95
96	EQUIPMENT	90
97	DRAINAGE	95
98	RETAIN. WALL	95
100	FOOTING	108*

Page	Item	Mod.
101	FOOTING	108*
	REBAR	111
102	EQUIPMENT	102*
	HAND DIG	108*
	REBAR	111
103	FOOTING	111*
	REBAR	111
104	PIER FOOTING	105*
105	ROUND PIER	105*
106	FTG./FOUND'N	105*
107	REBAR	111
	WHEEL/PUMP	111*
108	MONO SLAB	111*
109	PLAT/STEP	108*
110	SLABS	108*
111	CONCRETE	108*
	ASPHALT	103
112	BLOCK	98
113	BRICK	105
	BRICK/BLOCK	107
114	BLOCK	98
	BRICK	104
115	PLAT/STEP	97
116	CHIMNEY	107
	FLUE	97
	CONCR. FLUE	102*
117	FIREPLACE	102
	LINTEL	99
118	UNDERPIN	102*
119	RETAIN. WALL	107
	GRAVEL	102
120	BRICK	104
	FLAGSTONE	101
121	STUCCO	103*
	WATERPROOF	99
122	WOOD FOUND'N	93*
123	STEEL BEAM	101
	WOOD BEAM	99*
	LAMINATED	99*
	FLITCHES	98
	STEEL COL.	120
124	FLOOR FRAMING	98
125	I-JOISTS	99*
	PLYWOOD	118*
	PINE/FIR/HEM	118*
	STRAND BD.	116*
126	WOOD STUD	93*
	STEEL STUD	113
	FURRING	118*
127	PLYWOOD	118*
	ASPH.IMPR.	98
	FOAM FACE	98
	PINE	118*
	STRAND BD.	116*
128	ROOF FRAMING	99*
129	ROOF FRAMING	99*
	TRUSSES	99*
130	PLYWOOD	118*
	STRAND BD.	118*
	STRIP SHTH.	118*
132	ADJUSTMENTS	122
	ASPH/FIB	119
133	ROOFING	119
134	PLY ROOF	99
135	ADJUSTMENTS	122
	SLATE	107
136	ADJUSTMENTS	122

on Index

ovation Cost Estimator

Date: **Second Quarter 1995**

Area: **WASHINGTON, DC & VIC**

** Indicates change in % since last index*

Page	Item	Mod.
137	CEDAR	107*
138	ADJUSTMENTS	122
	PANELS	113*
139	PANELS	113*
140	ADJUSTMENTS	122
	CONCR	112
141	ADJUSTMENTS	122
	MIN. FIBER	113
142	TILE	110
143	COPPER	112
144	TEAR-OUT	95
145	PREPARATION	113*
146	REPAIR	122
	ALUM.	120
147	ASPH.	120
	REPAIR	122
148	ALUM.	112
	COPPER	114
	ITEMS	98
149	ALUM.	102
	GALV.	126
	COPPER	112
	WOOD	100*
	VINYL	100
	ITEMS	98
150	PINE/SPR/HEM	118*
	PLYWOOD	116*
	LOUVER	103
	LATTICE	97
151	ALUM.	87
	VINYL	100
152	PRESS TR.	98
	REDWOOD	102
	CEDAR	107*
153	PRESS TR.	98
	REDWOOD	102
	CEDAR	107*
154	PINE, FIR	124*
	ALUM.	87
	REDWOOD	112
	PLYWOOD	118*
	BEAD CEILING	124*
	FLOOR	99*
	SCREEN	98
	DOOR	99
155	IRON	99
	FIBERGL.	97
156	ADJUSTMENTS	95
	WRAP	98
	CEDAR	104
	REDWOOD	102
	FIR/PINE	118*
157	CEDAR	104
	REDWOOD	102
	FIR/PINE	118*
158	CEDAR SH.	113*
159	SHINGL PANEL	107*
160	HARDBOARD	96
161	HARDBOARD	96
	BATTENS	120*
162	CEDAR/RW	110
	FIR/PINE	120*
	PANELS	98
163	ALUM.	87
	VINYL	100
164	DOORS	99
165	DOORS	99
166	DOORS	99

Page	Item	Mod.
167	METAL DOOR	99
168	ENTRANCE	107
	WEATHERSTR.	98
	DEADBOLT	98
169	WOOD DOOR	107
	ALUM. DOOR	99
170	WOOD DOOR	106
	STEEL DOOR	99
	FIBER./ALUM.	99
	OPERATOR	98
171	ALUM. DOOR	98
	WOOD DOOR	106
172	PRE-HUNG	106
173	PRE-HUNG	106
174	FLUSH DOOR	104
175	SLIDING	106
176	DOORS	106
177	POCKET	104
178	BI-FOLD	106
179	BI-FOLD	106
180 THROUGH 189:		
	SINGL-GLAZE	103
	DOUBL-GLAZE	105
190	BAY	105
	FRAME	108*
	ROOF	120
	HEADBOARD	118*
191	BOW	105
	FRAME	108*
	ROOF	120
	HEADBOARD	118*
	GREENHOUSE	104
192	WINDOW	104
193	SKYLIGHT	104
194	DOUBLE	103
	SINGLE	102
195	SINGLE	103
	INSUL.	104
	AWNING	103
196	ADJUST	97
	WINDOWS	103
197	WINDOWS	104
198	WINDOWS	103
199	WINDOWS	104
200 THROUGH 206		99
207	REPLACE	99
208	NEW	101
	REPLACE	103
209	HW HEAT	101
	BASEBOARD	103
210	HW HEAT	101
211	FURNACE	103
212	HEAT	99
213	NEW SYST	120
	ON EXISTING	100
	WINDOW	99
214	DUCTS	97
	HUMIDIFIER	99
215	ELECTRIC	96
216	ELECTRIC	96
217	ELECTRIC	96
218	FANS, ETC.	97
219	FANS	97
	CENT. VAC.	98
	SMOKE DET.	97
220 THROUGH 223		112*
224	DRYWALL	104*
225	DRYWALL	104*

Page	Item	Mod.
226	GYP. LATH	112*
	METAL	102
	WOOD	124*
227	GYPSUM	103*
228	CERAMIC	101
229	PLYWOOD	98
	KN. PINE	124*
	CEDAR	113*
	REDWOOD	102
230	PATCH	106*
231	GYPSUM	103*
232	GYPSUM	103*
233	GYP. LATH	112*
	METAL	102
	WOOD	124*
234	PATCH	106*
235	COMP./GRID	98
	FURRING	102*
236	MOULDING	74
237	MOULDING	79
238	WOOD	79
	PLASTER	120
239	MILLWORK	74
	TRIM	79
240	#2 PINE	120*
	CWP	108
	OAK	101*
	PLYWOOD	112*
	ITEMS	79
241	#2 PINE	120*
	CWP	108
	OAK	101*
	PLYWOOD	112*
242	CLOSET	120*
243	ATTIC/CELLAR	124*
	MAIN	99
	IRON	98
244	SPIRAL	99
	FOLDING	98
245	ITEMS	98
246 THROUGH 253		99
254	FIXT./APPL.	99
255	ACCESSOR.	99
256	ACCESSOR.	99
257	ENCLOS.	98
258	SHUTTER	99
	FIREPLACE	99
259	SECURITY	98
260	FENCING	99
261	FENCING	98
262	OAK	101*
	FINISH	97
263	UNDERLAY	118*
	VINYL	100
264	UNDERLAY	118*
	TILE	101
	MARBLE	101
	SLATE	107
265	CARPET	98
266 THROUGH 284		108
285	CLEAN-UP	95
286	CLEAN-UP	95

SPECIFICATIONS		UNIT	PRICE	LOCAL AREA MODIFICATION %	LOCAL AREA MODIFICATION PRICE
HARDBOARD PANEL SIDING	• 4'-0" WIDTH PANELS • 7/16" THICKNESS • PRIMED • ROUGH SAWN TEXTURE • #15 FELT PAPER • 3/4" X 4" OUTSIDE CORNERS • 3/4" X 3/4" INSIDE CORNERS				
	8' LENGTH	SF	2.54		
	9' LENGTH	SF	2.60		
	SAME AS ABOVE, PREFINISHED				
	8' LENGTH	SF	2.78		
	9' LENGTH	SF	2.79		
HARDBOARD IMITATION STUCCO SIDING	• 4'-0" WIDTH PANELS • 7/16" THICKNESS • PRIMED • STUCCO TEXTURE • #15 FELT PAPER • 3/4" X 4" OUTSIDE CORNERS • 3/4" X 3/4" INSIDE CORNERS				
	8' LENGTH	SF	3.00		
	9' LENGTH	SF	3.05		
	SAME AS ABOVE, PREFINISHED				
	8' LENGTH	SF	3.71		
	9' LENGTH	SF	3.75		
BATTENS	INSTALL BATTENS ON VERTICAL JOINTS OR TO ACHIEVE TUDOR EFFECT				
	1" X 4" BATTENS	LF	1.44		
	1" X 6" BATTENS	LF	1.86		
	LF = TOTAL LENGTH OF BATTENS				

11. PLYWOOD SIDING PANELS

SPECIFICATIONS		UNIT	PRICE	LOCAL AREA MODIFICATION	
				%	PRICE
TEXTURE 1-11	• PLYWOOD PANELS WITH PATTERNS 4", 6" OR 8" O.C. • #15 FELT PAPER • 1" X 3" AND 1" X 4" TRIM • RUST RESISTANT NAILS				
	Cedar 3/8"	SF	2.96		
	5/8"	SF	3.48		
	Redwood 3/8"	SF	3.17		
	5/8"	SF	3.56		
	Fir 3/8"	SF	1.98		
	5/8"	SF	2.28		
	Pine 3/8"	SF	1.98		
	5/8"	SF	2.31		
NATURAL STONE CAST PANELS	• NATURAL STONE CHIPS BONDED TO PLY-WOOD PANELS • RUST RESISTANT OR COLOR NAILS • #15 FELT PAPER • SEALANTS AND CAULKING AS REQUIRED				
	48" X 1/2" X 8'	SF	6.20		
	48" X 1/2" X 9', 10'	SF	6.65		

11

SPECIFICATIONS		UNIT	PRICE	LOCAL AREA MODIFICATION	
				%	PRICE
ALUMINUM SIDING	• .024" THICK HORIZONTAL ALUMINUM SIDING • INSULATION BOARD BACKING • #15 FELT PAPER • ALL ACCESSORIES, BUT **NOT** INCLUDING SOFFIT, FASCIA OR WINDOW AND DOOR TRIM				
	8" SMOOTH	SF	4.08		
	8" DOUBLE 4	SF	4.13		
	10" SMOOTH	SF	4.11		
	10" DOUBLE 5	SF	4.28		
	SAME AS ABOVE, 12" ALUMINUM VERTICAL BOARD AND BATTEN	SF	4.02		
VINYL SIDING	• SOLID HORIZONTAL PVC VINYL SIDING • INSULATION BOARD BACKING • #15 FELT PAPER • ALL ACCESSORIES, BUT **NOT** INCLUDING SOFFIT, FASCIA OR WINDOW AND DOOR TRIM				
	8" SMOOTH	SF	3.41		
	8" DOUBLE 4	SF	3.54		
	10" SMOOTH	SF	3.47		
	10" DOUBLE 5	SF	3.63		
	SAME AS ABOVE, 12" VINYL VERTICAL BOARD AND BATTEN	SF	3.41		
WITHOUT INSULATION	WHEN INSULATION BOARD BACKING IS NOT USED				
	DEDUCT	SF	.47		

12. EXTERIOR DOORS

SPECIFICATIONS		UNIT	PRICE	LOCAL AREA MODIFICATION %	LOCAL AREA MODIFICATION PRICE
	NOTE: ALL EXTERIOR DOORS ARE 1-3/4" THICK				
FLUSH, BIRCH	PARTICLE CORE, INCLUDING FRAME, BRICK MOULDING, 1-1/2 PAIR 4 X 4 BUTTS, INTERIOR CASINGS, ENTRANCE LOCK @ $20.00				
	2-8 X 6-8	EA	504.00		
	3-0 X 6-8	EA	519.00		
	3-0 X 7-0	EA	537.00		
	TO ABOVE DOORS, FOR EACH LIGHT **ADD**	EA	13.05		
COLONIAL	6 PANEL INCLUDING FRAME, TRIM AND HARDWARE, FIR				
	3-0 X 6-8	EA	696.00		
	3-0 X 7-0	EA	717.00		
	SAME AS ABOVE, PINE DOOR				
	3-0 X 6-8	EA	696.00		
	3-0 X 7-0	EA	816.00		
	SAME AS ABOVE, OAK DOOR				
	3-0 X 6-8	EA	849.00		
	3-0 X 7-0	EA	900.00		
COLONIAL WITH LIGHTS	ABOVE COLONIAL DOORS WITH 4 SMALL LIGHTS AT TOP **ADD**	EA	22.50		
DOUBLE DOORS	2 DOORS, INCLUDING FRAME, BRICK MOULDING, SILL, 3 PAIR 4 X 4 BUTTS, KEYLOCK, BARREL BOLT, T-ASTRAGEL, CASINGS, 6-0 X 6-8				
	FLUSH BIRCH	SET	954.00		
	COLONIAL FIR	SET	1302.00		
	COLONIAL PINE	SET	1458.00		
	COLONIAL OAK	SET	1611.00		
SIDE OR REAR DOOR	FIR, 3 PANELS, 4 LIGHTS, FRAME, TRIM, BRICK MOULDING, HARDWARE, 2-8 X 6-8	EA	612.00		
REPLACE EXTERIOR DOOR	REMOVE EXTERIOR DOOR AND REPLACE WITH 6 PANEL COLONIAL FIR IN EXISTING FRAME AND CASING, INSTALL EXISTING LOCKSET	EA	501.00		

12

SPECIFICATIONS		UNIT	PRICE	LOCAL AREA MODIFICATION	
				%	PRICE
FRENCH	FIR, 1-3/4", 10 LIGHTS, INCLUDING FRAME, BRICK MOULDING, OAK SILL, 1-1/2 PAIR 4 X 4 BUTTS, KEY-LOCK AND INTERIOR CASINGS				
	2-6 X 6-8	EA	648.00		
	2-8 X 6-8	EA	678.00		
	3-0 X 6-8	EA	702.00		
	2 DOORS, INCLUDING FRAME, BRICK MOULDING, SILL, 3 PAIR 4 X 4 BUTTS, KEYLOCK, BARREL BOLT, T-ASTRAGEL, CASINGS				
	5-0 X 6-8	SET	1149.00		
	5-4 X 6-8	SET	1323.00		
	6-0 X 6-8	SET	1338.00		
DUTCH	FIR, 2 PANEL, 9 LIGHTS, WITH FRAME, BRICK MOULDING AND OAK SILL, 2 PAIR 4 X 4 BUTTS, KEYLOCK, DUTCH DOOR QUADRANT, INTERIOR CASINGS				
	2-8 X 6-8	EA	870.00		
	3-0 X 6-8	EA	915.00		
	FOR CROSSBUCK STYLE **ADD**	EA	34.50		

12. PRE-HUNG EXTERIOR DOORS

SPECIFICATIONS		UNIT	PRICE	LOCAL AREA MODIFICATION %	LOCAL AREA MODIFICATION PRICE
FLUSH, BIRCH	1-3/4", SOLID CORE, INCLUDING FRAME, BRICK MOULDING, THRESHOLD, INTERIOR CASINGS, ENTRANCE LOCK @ $20.00				
	2-8 X 6-8	EA	387.00		
	3-0 X 6-8	EA	405.00		
	3-0 X 7-0	EA	447.00		
COLONIAL	6 PANEL, INCLUDING FRAME, TRIM AND HARD-WARE, FIR				
	3-0 X 6-8	EA	594.00		
	3-0 X 7-0	EA	648.00		
	SAME AS ABOVE, PINE DOOR				
	3-0 X 6-8	EA	666.00		
	3-0 X 7-0	EA	738.00		
	SAME AS ABOVE, OAK DOOR				
	3-0 X 6-8	EA	777.00		
	3-0 X 7-0	EA	822.00		

12. METAL PRE-HUNG EXTERIOR DOORS

SPECIFICATIONS		UNIT	PRICE	LOCAL AREA MODIFICATION	
				%	PRICE
METAL PRE-HUNG	• 6 PANEL • 3-0 X 6-8 X 1-3/4" • FOAM CORE • 1-1/2 PAIR BUTTS • ALUMINUM THRESHOLD • MAGNETIC WEATHERSTRIPPING • ENTRANCE LOCK • EXTERIOR FRAME AND BRICK MOULDING • INTERIOR CASINGS	EA	399.00		

12

12. DOOR ACCESSORIES

SPECIFICATIONS		UNIT	PRICE	LOCAL AREA MODIFICATION %	LOCAL AREA MODIFICATION PRICE
	NOTE: **ADD** ORNAMENTAL ENTRANCE FRAME AND SIDELIGHT COSTS TO MATERIALS AND LABOR COSTS OF DOOR SELECTED.				
ORNAMENTAL ENTRANCE FRAME & SIDELIGHTS	• ENTRANCE FRAME WITH 2 PILASTERS AND 11" HIGH PEDIMENT • SIDELIGHT, 1-2 X 6-8 X 1-3/4" WITH 7/16" INSULATED GLASS, ONE LIGHT OR ONE LIGHT AND ONE PANEL				
	SINGLE DOOR **ADD**	EA	204.00		
	SINGLE DOOR WITH 1 SIDELIGHT **ADD**	EA	591.00		
	SINGLE DOOR WITH 2 SIDELIGHTS **ADD**	EA	969.00		
	• ENTRANCE FRAME WITH 2 PILASTERS AND 18" HIGH PEDIMENT WITH SLOPED OR ARCHED SIDES				
	SINGLE DOOR **ADD**	EA	282.00		
	SINGLE DOOR WITH 1 SIDELIGHT **ADD**	EA	678.00		
	SINGLE DOOR WITH 2 SIDELIGHTS **ADD**	EA	1041.00		
	3 OR 5 LIGHT DIVIDED LIGHT GRILLE **ADD**	EA	39.00		
DOOR WEATHER-STRIPPING	ALUMINUM THRESHOLD AND BRONZE WEATHER-STRIPPING	EA	78.00		
	INTERLOCKING WEATHERSTRIPPING ON 3-0 X 6-8 DOOR ON WOOD FLOOR, INCLUDING THRESHOLD	EA	153.00		
DEADBOLT LOCK	OUTSIDE KEY, THUMB TURN INSIDE, SINGLE CYLINDER				
	STANDARD QUALITY	EA	60.00		
	PREMIUM QUALITY	EA	108.00		

12

SPECIFICATIONS			UNIT	PRICE	LOCAL AREA MODIFICATION %	LOCAL AREA MODIFICATION PRICE
GLIDING PATIO DOOR, WOOD	WITH INSULATED GLASS, INCLUDING FRAME, HARDWARE AND INTERIOR AND EXTERIOR TRIM					
	Panels	***Economy***				
	2	6-0 X 6-8	EA	1044.00		
	2	8-0 X 6-8	EA	1233.00		
	3	9-0 X 6-8	EA	1470.00		
	3	12-0 X 6-8	EA	1860.00		
		Builder				
	2	6-0 X 6-8	EA	1194.00		
	2	8-0 X 6-8	EA	1407.00		
	3	9-0 X 6-8	EA	1635.00		
	3	12-0 X 6-8	EA	2061.00		
		Premium				
	2	6-0 X 6-8	EA	1530.00		
	2	8-0 X 6-8	EA	1752.00		
	3	9-0 X 6-8	EA	1983.00		
	3	12-0 X 6-8	EA	2601.00		
ATRIUM STYLE WOOD PATIO DOOR	ATRIUM STYLE WITH HINGED ACTIVE SIDE, INSULATED GLASS, FRAME, HARDWARE & EXTERIOR & INTERIOR TRIM					
	Panels	***Economy***				
	2	6-0 X 6-8	EA	1005.00		
	2	8-0 X 6-8	EA	1110.00		
		Builder				
	2	6-0 X 6-8	EA	1152.00		
	2	8-0 X 6-8	EA	1275.00		
		Premium				
	2	6-0 X 6-8	EA	1449.00		
	2	8-0 X 6-8	EA	1584.00		
GLIDING PATIO DOOR, ALUMINUM	WITH INSULATED GLASS, INCLUDING FRAME, HARDWARE AND INTERIOR AND EXTERIOR TRIM					
	Panels	***Opening***				
	2	5-0 X 6-8	EA	672.00		
	2	6-0 X 6-8	EA	735.00		
	2	8-0 X 6-8	EA	891.00		
	2	9-0 X 6-8	EA	1017.00		
	2	12-0 X 6-8	EA	1086.00		
	2	15-0 X 6-8	EA	1161.00		

12. GARAGE DOORS

SPECIFICATIONS		UNIT	PRICE	LOCAL AREA MODIFICATION	
				%	PRICE
GARAGE DOOR, WOOD	4 SECTION 4 PANEL, ONE ROW OF LIGHTS (OPTIONAL) INCLUDING STOPS, HARDWARE AND EXTERIOR TRIM				
	8-0 X 7-0	EA	600.00		
	9-0 X 7-0	EA	639.00		
	16-0 X 7-0	EA	1101.00		
GARAGE DOOR, METAL	SAME AS ABOVE, STEEL GARAGE DOOR				
	8-0 X 7-0	EA	627.00		
	9-0 X 7-0	EA	651.00		
	16-0 X 7-0	EA	1107.00		
	SAME AS ABOVE, FIBERGLASS WITH ALUMINUM FRAME				
	8-0 X 7-0	EA	681.00		
	9-0 X 7-0	EA	723.00		
	16-0 X 7-0	EA	1239.00		
GARAGE DOOR OPERATOR	AUTOMATIC CONTROL, 1/4 HP REVERSIBLE MOTOR, CHAIN DRIVE, FOR DOOR UP TO 16 X 7				
	ONE CAR SET	EA	441.00		
	TWO CAR SET	EA	507.00		
	SAME AS ABOVE WITH 1/3 HP MOTOR FOR DOOR UP TO 18 FEET				
	ONE CAR SET	EA	498.00		
	TWO CAR SET	EA	561.00		

12. STORM/SCREEN COMBINATION DOORS

SPECIFICATIONS		UNIT	PRICE	LOCAL AREA MODIFICATION	
				%	PRICE
ALUMINUM STORM/ SCREEN COMBINATION DOORS	1" ALUMINUM FRAME WITH TEMPERED GLASS & SCREEN INSERTS, WEATHERSTRIPPED, KICK PANEL, ALL HARDWARE, CLOSER, 6-8 OR 7-0 HIGH X 2-6, 2-8, 2-10 OR 3-0 WIDE, 2 GLASS INSERTS AND 2 SCREEN INSERTS				
	ANODIZED	EA	258.00		
	WHITE	EA	270.00		
	SELF-STORING DOOR WITH 2 GLASS INSERTS AND 1 SCREEN INSERT				
	ECONOMY ANODIZED	EA	261.00		
	PREMIUM WHITE	EA	297.00		
	CROSSBUCK, WITH 1 TEMPERED GLASS AND 1 UPPER SCREEN INSERT, COLONIAL FRINGE, 3 BLACK STRAP HINGES, BLACK HANDLE, CROSSBUCK KICK PANEL				
	ECONOMY WHITE	EA	288.00		
	PREMIUM WHITE	EA	324.00		
WOOD STORM/ SCREEN COMBINATION DOORS	1-1/8" THICK WOOD FRAME COMBINATION DOOR WITH 1 GLASS INSERT, INCLUDING CLOSER AND ALL HARDWARE, 6-8 AND 7-0 HIGH X 2-6, 2-8 AND 3-0 WIDE, CUT OR PLANE TO FIT	EA	348.00		

12. PRE-HUNG INTERIOR DOOR UNITS

SPECIFICATIONS		UNIT	PRICE	LOCAL AREA MODIFICATION	
				%	PRICE
PRE-HUNG DOOR	• INTERIOR 1-3/8" DOOR • STAIN GRADE JAMB • 2 SIDES CASING • PRIVACY LOCK @ $10				
	BIRCH FLUSH HOLLOW CORE				
	1-6 X 6-8	EA	192.00		
	2-0 X 6-8	EA	193.50		
	2-4 X 6-8	EA	195.00		
	2-6 X 6-8	EA	198.00		
	2-8 X 6-8	EA	204.00		
	3-0 X 6-8	EA	210.00		
	OAK FLUSH HOLLOW CORE				
	1-6 X 6-8	EA	225.00		
	2-0 X 6-8	EA	228.00		
	2-4 X 6-8	EA	237.00		
	2-6 X 6-8	EA	240.00		
	2-8 X 6-8	EA	246.00		
	3-0 X 6-8	EA	252.00		
	WALNUT FL. HOLLOW CORE				
	1-6 X 6-8	EA	330.00		
	2-0 X 6-8	EA	339.00		
	2-4 X 6-8	EA	345.00		
	2-6 X 6-8	EA	348.00		
	2-8 X 6-8	EA	351.00		
	3-0 X 6-8	EA	357.00		
	PINE, 6 PANEL				
	1-6 X 6-8	EA	294.00		
	2-0 X 6-8	EA	309.00		
	2-4 X 6-8	EA	321.00		
	2-6 X 6-8	EA	324.00		
	2-8 X 6-8	EA	330.00		
	3-0 X 6-8	EA	348.00		
	DOUBLE FRENCH DOORS, 15 LITE				
	1-6 X 6-8	SET	678.00		
	2-0 X 6-8	SET	714.00		
	2-4 X 6-8	SET	747.00		
	2-6 X 6-8	SET	828.00		
	2-8 X 6-8	SET	912.00		
	3-0 X 6-8	SET	960.00		

12. PRE-HUNG INTERIOR DOOR UNITS

SPECIFICATIONS		UNIT	PRICE	LOCAL AREA MODIFICATION	
				%	PRICE
PRE-HUNG DOOR	• INTERIOR 1-3/8" DOOR • PAINT GRADE JAMB • 2 SIDES CASING • PRIVACY LOCK @ $10				
	MAHOGANY FLUSH HC				
	1-6 X 6-8	EA	181.50		
	2-0 X 6-8	EA	183.00		
	2-4 X 6-8	EA	186.00		
	2-6 X 6-8	EA	189.00		
	2-8 X 6-8	EA	192.00		
	3-0 X 6-8	EA	198.00		
	PINE LOUVER / LOUVER				
	1-6 X 6-8	EA	219.00		
	2-0 X 6-8	EA	228.00		
	2-4 X 6-8	EA	240.00		
	2-6 X 6-8	EA	243.00		
	2-8 X 6-8	EA	258.00		
	3-0 X 6-8	EA	276.00		
	PINE LOUVER / PANEL				
	1-6 X 6-8	EA	228.00		
	2-0 X 6-8	EA	243.00		
	2-4 X 6-8	EA	252.00		
	2-6 X 6-8	EA	255.00		
	2-8 X 6-8	EA	273.00		
	3-0 X 6-8	EA	288.00		
	HARDBOARD FLUSH HC				
	1-6 X 6-8	EA	162.00		
	2-0 X 6-8	EA	168.00		
	2-4 X 6-8	EA	171.00		
	2-6 X 6-8	EA	174.00		
	2-8 X 6-8	EA	177.00		
	3-0 X 6-8	EA	180.00		
	6 RAISED PANELS, HARD-BOARD FACED				
	1-6 X 6-8	EA	195.00		
	2-0 X 6-8	EA	201.00		
	2-4 X 6-8	EA	204.00		
	2-6 X 6-8	EA	210.00		
	2-8 X 6-8	EA	213.00		
	3-0 X 6-8	EA	219.00		

12. INTERIOR FLUSH DOORS

SPECIFICATIONS		UNIT	PRICE	LOCAL AREA MODIFICATION	
				%	PRICE
FLUSH	MAHOGANY, HOLLOW CORE, 1-3/8", INCLUDING JAMBS AND CASINGS, 1 PAIR 3-1/2 X 3-1/2 BUTTS, PRIVACY LOCK				
	1-4 X 6-8	EA	228.00		
	1-6 X 6-8	EA	228.00		
	1-8 X 6-8	EA	231.00		
	2-0 X 6-8	EA	231.00		
	2-4 X 6-8	EA	231.00		
	2-6 X 6-8	EA	237.00		
	2-8 X 6-8	EA	240.00		
	3-0 X 6-8	EA	246.00		
	BIRCH, HOLLOW CORE, 1-3/8", INCLUDING TRIM AND HARDWARE				
	1-4 X 6-8	EA	234.00		
	1-6 X 6-8	EA	234.00		
	1-8 X 6-8	EA	240.00		
	2-0 X 6-8	EA	240.00		
	2-4 X 6-8	EA	240.00		
	2-6 X 6-8	EA	252.00		
	2-8 X 6-8	EA	255.00		
	3-0 X 6-8	EA	267.00		
	HARDBOARD, UNFINISHED HOLLOW CORE, 1-3/8" DOOR, INCLUDING TRIM AND HARDWARE				
	1-4 X 6-8	EA	216.00		
	1-6 X 6-8	EA	216.00		
	1-8 X 6-8	EA	222.00		
	2-0 X 6-8	EA	222.00		
	2-4 X 6-8	EA	222.00		
	2-6 X 6-8	EA	225.00		
	2-8 X 6-8	EA	228.00		
	3-0 X 6-8	EA	228.00		

SPECIFICATIONS		UNIT	PRICE	LOCAL AREA MODIFICATION	
				%	PRICE
SLIDING	MAHOGANY FLUSH HOLLOW CORE, 1-3/8", 2 DOORS, INCLUDING JAMBS, TWO SIDES 2-1/4" CASING, ALL NECESSARY HARDWARE				
	4-0 X 6-8	EA	276.00		
	5-0 X 6-8	EA	291.00		
	6-0 X 6-8	EA	303.00		
	SAME AS ABOVE WITH BIRCH HOLLOW CORE				
	4-0 X 6-8	EA	303.00		
	5-0 X 6-8	EA	321.00		
	6-0 X 6-8	EA	339.00		
	SAME AS ABOVE WITH PINE, HEMLOCK, OR FIR LOUVER DOORS				
	4-0 X 6-8	EA	426.00		
	5-0 X 6-8	EA	465.00		
	6-0 X 6-8	EA	525.00		
	SAME AS ABOVE WITH OAK HOLLOW CORE				
	4-0 X 6-8	EA	333.00		
	5-0 X 6-8	EA	348.00		
	6-0 X 6-8	EA	369.00		

12. INTERIOR DOORS

SPECIFICATIONS		UNIT	PRICE	LOCAL AREA MODIFICATION	
				%	PRICE
COLONIAL	HARDBOARD FACED, 6 RAISED PANELS, 1-3/8", INCLUDING JAMBS, STOPS, 2 SIDES 2-1/2" CASING, 1 PAIR 3-1/2 X 3-1/2 BUTTS, PRIVACY LOCK				
	1-6 X 6-8	EA	246.00		
	2-0 X 6-8	EA	261.00		
	2-2 X 6-8	EA	264.00		
	2-4 X 6-8	EA	267.00		
	2-6 X 6-8	EA	276.00		
	2-8 X 6-8	EA	282.00		
	3-0 X 6-8	EA	285.00		
	SAME AS ABOVE, WITH SOLID PINE RAISED PANELS				
	1-6 X 6-8	EA	321.00		
	2-0 X 6-8	EA	351.00		
	2-2 X 6-8	EA	351.00		
	2-4 X 6-8	EA	360.00		
	2-6 X 6-8	EA	369.00		
	2-8 X 6-8	EA	381.00		
	3-0 X 6-8	EA	399.00		
LOUVERED	PINE, 1-3/8", WITH OR WITHOUT CROSS RAIL, INCLUDING JAMBS, STOPS, 2 SIDES 2-1/4" CASING, 1 PAIR 3-1/2 X 3-1/2 BUTTS, PASSAGE SET, 1/4 X 1-3/8 SLATS				
	1-6 X 6-8	EA	261.00		
	2-0 X 6-8	EA	291.00		
	2-4 X 6-8	EA	300.00		
	2-6 X 6-8	EA	303.00		
	2-8 X 6-8	EA	309.00		
	3-0 X 6-8	EA	318.00		

SPECIFICATIONS		UNIT	PRICE	LOCAL AREA MODIFICATION %	LOCAL AREA MODIFICATION PRICE
POCKET DOOR	METAL REINFORCED SLIDING DOOR POCKET, INCLUDING TRACK, NYLON ROLLERS, JAMBS, CASINGS AND HARDWARE, 1-3/8" DOOR				
	BIRCH FLUSH HOLLOW CORE				
	2-0 X 6-8	EA	381.00		
	2-4 X 6-8	EA	381.00		
	2-6 X 6-8	EA	396.00		
	2-8 X 6-8	EA	402.00		
	3-0 X 6-8	EA	414.00		
	HARDBOARD FLUSH HC				
	2-0 X 6-8	EA	366.00		
	2-4 X 6-8	EA	366.00		
	2-6 X 6-8	EA	372.00		
	2-8 X 6-8	EA	378.00		
	3-0 X 6-8	EA	384.00		
	COLONIAL WITH SOLID RAISED PANELS				
	2-0 X 6-8	EA	447.00		
	2-4 X 6-8	EA	450.00		
	2-6 X 6-8	EA	465.00		
	2-8 X 6-8	EA	483.00		
	3-0 X 6-8	EA	498.00		

12. BI-FOLD DOORS

SPECIFICATIONS		UNIT	PRICE	LOCAL AREA MODIFICATION %	LOCAL AREA MODIFICATION PRICE
BIRCH BI-FOLD	BIRCH HOLLOW CORE FLUSH 1-3/8" DOORS, INCLUDING JAMBS, 2 SIDES 2-1/4" CASING, BI-FOLD HARDWARE				
	Opening Size				
	TWO DOORS 2-6 X 6-8	SET	204.00		
	3-0 X 6-8	SET	216.00		
	FOUR DOORS 4-0 X 6-8	SET	255.00		
	5-0 X 6-8	SET	270.00		
	6-0 X 6-8	SET	285.00		
MAHOGANY BI-FOLD	SAME AS ABOVE WITH MAHOGANY HOLLOW CORE				
	TWO DOORS 2-6 X 6-8	SET	201.00		
	3-0 X 6-8	SET	213.00		
	FOUR DOORS 4-0 X 6-8	SET	249.00		
	5-0 X 6-8	SET	267.00		
	6-0 X 6-8	SET	282.00		
OAK BI-FOLD	SAME AS ABOVE WITH OAK HOLLOW CORE				
	TWO DOORS 2-6 X 6-8	SET	216.00		
	3-0 X 6-8	SET	234.00		
	FOUR DOORS 4-0 X 6-8	SET	273.00		
	5-0 X 6-8	SET	291.00		
	6-0 X 6-8	SET	309.00		
HARDBOARD BI-FOLD	SAME AS ABOVE WITH HARDBOARD HOLLOW CORE				
	TWO DOORS 2-6 X 6-8	SET	189.00		
	3-0 X 6-8	SET	198.00		
	FOUR DOORS 4-0 X 6-8	SET	234.00		
	5-0 X 6-8	SET	246.00		
	6-0 X 6-8	SET	264.00		

SPECIFICATIONS		UNIT	PRICE	LOCAL AREA MODIFICATION	
				%	PRICE
COLONIAL BI-FOLD	BIRCH HOLLOW CORE FLUSH 1-3/8" DOORS, INCLUDING JAMBS, 2 SIDES 2-1/4" CASING, BI-FOLD HARDWARE				
	Opening Size				
	TWO DOORS 2-6 X 6-8	SET	234.00		
	3-0 X 6-8	SET	246.00		
	FOUR DOORS 4-0 X 6-8	SET	294.00		
	5-0 X 6-8	SET	312.00		
	6-0 X 6-8	SET	330.00		
PINE LOUVERED BI-FOLD	SAME AS ABOVE WITH PINE LOUVERED DOORS				
	TWO DOORS 2-6 X 6-8	SET	261.00		
	3-0 X 6-8	SET	270.00		
	FOUR DOORS 4-0 X 6-8	SET	327.00		
	5-0 X 6-8	SET	345.00		
	6-0 X 6-8	SET	366.00		
MIRROR BI-FOLD	SAME AS ABOVE WITH FRAMELESS MIRROR WITH SHATTER-RESISTANT SAFETY BACKING				
	TWO DOORS 2-6 X 6-8	SET	282.00		
	3-0 X 6-8	SET	318.00		
	FOUR DOORS 4-0 X 6-8	SET	393.00		
	5-0 X 6-8	SET	426.00		
	6-0 X 6-8	SET	498.00		

13. SINGLE GLAZED WINDOWS

SPECIFICATIONS			UNIT	PRICE	LOCAL AREA MODIFICATION %	LOCAL AREA MODIFICATION PRICE
	NOTE: ALL *WINDOW SIZES* SHOWN ON THIS AND THE FOLLOWING PAGES ARE *SASH SIZES* (1-8 X 3-2 = 20" WIDE X 38" HIGH)					
WOOD DOUBLE-HUNG	• FRAME & BRICK MOULD • SINGLE GLAZED SASH • SASH LOCK • INTERIOR TRIM, STOPS • STOOL AND APRON					
	2 LIGHTS	1-8 X 3-2	EA	171.00		
		3-10	EA	177.00		
		4-2	EA	186.00		
		4-6	EA	192.00		
	4 LIGHTS	1-8 X 3-2	EA	192.00		
		3-10	EA	195.00		
		4-2	EA	201.00		
		4-6	EA	207.00		
	6 LIGHTS	2-0 X 3-2	EA	198.00		
		3-10	EA	201.00		
		4-2	EA	207.00		
		4-6	EA	210.00		
		2-4 X 3-2	EA	201.00		
		3-10	EA	207.00		
		4-2	EA	210.00		
		4-6	EA	222.00		
		2-8 X 3-2	EA	210.00		
		3-10	EA	216.00		
		4-2	EA	222.00		
		4-6	EA	237.00		
		5-2	EA	258.00		
		5-6	EA	270.00		
	8 LIGHTS	3-0 X 3-2	EA	216.00		
		3-10	EA	222.00		
		4-2	EA	243.00		
		4-6	EA	252.00		
		5-2	EA	273.00		
		5-6	EA	282.00		
TWIN UNITS	DOUBLE THE SINGLE UNIT AMOUNT AND **ADD**		EA	24.00		
TRIPLE UNITS	TRIPLE THE SINGLE UNIT AMOUNT AND **ADD**		EA	33.00		

13

SPECIFICATIONS		UNIT	PRICE	LOCAL AREA MODIFICATION %	LOCAL AREA MODIFICATION PRICE
ONE LIGHT PICTURE SASH AND FRAME	WOOD, WITH 1/4" PLATE GLASS, INCLUDING EXTERIOR BRICK MOULD, FRAME, SASH, INTERIOR CASINGS, STOPS, STOOL				
	4-0 X 3-10	EA	345.00		
	5-0 X 3-10	EA	393.00		
	6-0 X 3-10	EA	426.00		
	4-0 X 4-2	EA	360.00		
	5-0 X 4-2	EA	408.00		
	6-0 X 4-2	EA	480.00		
	4-0 X 4-6	EA	390.00		
	5-0 X 4-6	EA	447.00		
	6-0 X 4-6	EA	465.00		
	4-0 X 5-2	EA	426.00		
	5-0 X 5-2	EA	474.00		
	6-0 X 5-2	EA	546.00		
	SAME AS ABOVE WITH 1/2" INSULATED GLASS PICTURE SASH **ADD**	EA	50%		
PICTURE WINDOW	SINGLE GLAZED PICTURE WINDOW WITH FLANKING DOUBLE-HUNG FRAMES AND SASH				
	Side Center Side Height				
	2-0 4-0 2-0 4-6	EA	681.00		
	5-0	EA	735.00		
	6-0	EA	759.00		
	3-0 4-0 3-0 5-2	EA	846.00		
	5-0	EA	897.00		
	6-0	EA	969.00		
	SAME AS ABOVE USING 1/2" INSULATED GLASS IN CENTER SECTION **ADD**	EA	60%		

13

13. DOUBLE GLAZED WINDOWS -- BUILDER QUALITY

SPECIFICATIONS				UNIT	PRICE	LOCAL AREA MODIFICATION %	LOCAL AREA MODIFICATION PRICE
WOOD DOUBLE HUNG	• FRAME, SASH, BRICK MOULD • PRIMED SASH WITH 1/2" INSULATING GLASS • SASH LOCK • INTERIOR TRIM, STOPS • STOOL AND APRON • SCREEN						
	2 LIGHTS	1-8 X	4-2	EA	324.00		
			4-6	EA	324.00		
		2-0 X	3-2	EA	306.00		
			4-2	EA	321.00		
			4-6	EA	336.00		
			5-2	EA	342.00		
		2-4 X	2-10	EA	297.00		
			3-2	EA	300.00		
			3-10	EA	318.00		
			4-2	EA	327.00		
			4-6	EA	333.00		
			5-2	EA	348.00		
		2-8 X	2-10	EA	327.00		
			3-2	EA	333.00		
			3-10	EA	348.00		
			4-2	EA	363.00		
			4-6	EA	378.00		
			5-2	EA	390.00		
			5-6	EA	405.00		
			6-2	EA	420.00		
		3-0 X	2-10	EA	357.00		
			3-2	EA	369.00		
			3-10	EA	381.00		
			4-2	EA	396.00		
			4-6	EA	405.00		
			5-2	EA	420.00		
			5-6	EA	438.00		
			6-2	EA	450.00		
		3-4 X	3-2	EA	396.00		
			3-10	EA	414.00		
			4-2	EA	429.00		
			4-6	EA	444.00		
			5-2	EA	459.00		
			5-6	EA	474.00		
		3-8 X	4-6	EA	462.00		
TWIN UNITS	DOUBLE THE SINGLE UNIT AMOUNT AND **ADD**			EA	24.00		
TRIPLE UNITS	TRIPLE THE SINGLE UNIT AMOUNT AND **ADD**			EA	33.00		

13

13. DOUBLE GLAZED WINDOWS -- BUILDER QUALITY

SPECIFICATIONS		UNIT	PRICE	LOCAL AREA MODIFICATION %	LOCAL AREA MODIFICATION PRICE
DIVIDED LIGHT GRILLE	DIVIDED LIGHT GRILLES FOR WINDOWS **ADD**	SF	3.45		
SCREENS	FULL LENGTH INSECT SCREENS FOR WINDOWS **ADD**	SF	2.60		
ONE LIGHT PICTURE SASH & FRAME	• FRAME, SASH AND BRICK MOULDING • PRIMED SASH WITH 1/2" INSULATING GLASS • INTERIOR TRIM, STOPS, STOOL AND APRON				
	3-0 X 4-2	EA	423.00		
	4-6	EA	447.00		
	4-0 X 4-2	EA	471.00		
	4-6	EA	492.00		
	5-2	EA	540.00		
	5-4 X 4-6	EA	585.00		
PICTURE WINDOW WITH FLANKERS	PICTURE WINDOW ABOVE WITH DOUBLE HUNG FRAMES AND SASH, INCLUDING ALL TRIM, ETC. TOTAL THE COSTS FOR DOUBLE HUNG AND PICTURE WINDOW AND **ADD** EA = TOTAL UNIT	EA	16.50		

13

13. DOUBLE GLAZED WINDOWS -- PREMIUM QUALITY

SPECIFICATIONS				UNIT	PRICE	LOCAL AREA MODIFICATION %	LOCAL AREA MODIFICATION PRICE
WOOD DOUBLE HUNG	• VINYL COVERED FRAME AND BRICK MOULDING • PRIMED SASH WITH 1/2" INSULATING GLASS • SASH LOCK • INTERIOR TRIM, STOPS • STOOL AND APRON • SCREEN						
	2 LIGHTS	1-8 X	4-2	EA	387.00		
			4-6	EA	387.00		
		2-0 X	3-2	EA	351.00		
			4-2	EA	354.00		
			4-6	EA	375.00		
			5-2	EA	387.00		
		2-4 X	2-10	EA	369.00		
			3-2	EA	390.00		
			3-10	EA	411.00		
			4-2	EA	420.00		
			4-6	EA	429.00		
			5-2	EA	432.00		
		2-8 X	2-10	EA	390.00		
			3-2	EA	399.00		
			3-10	EA	417.00		
			4-2	EA	441.00		
			4-6	EA	453.00		
			5-2	EA	471.00		
			5-6	EA	489.00		
			6-2	EA	507.00		
		3-0 X	2-10	EA	429.00		
			3-2	EA	444.00		
			3-10	EA	456.00		
			4-2	EA	480.00		
			4-6	EA	489.00		
			5-2	EA	507.00		
			5-6	EA	528.00		
			6-2	EA	546.00		
		3-4 X	3-2	EA	477.00		
			3-10	EA	495.00		
			4-2	EA	519.00		
			4-6	EA	534.00		
			5-2	EA	555.00		
			5-6	EA	573.00		
		3-8 X	4-6	EA	561.00		
TWIN UNITS	DOUBLE THE SINGLE UNIT AMOUNT AND **ADD**			EA	24.00		
TRIPLE UNITS	TRIPLE THE SINGLE UNIT AMOUNT AND **ADD**			EA	36.00		

13

13. DOUBLE GLAZED WINDOWS -- PREMIUM QUALITY

SPECIFICATIONS		UNIT	PRICE	LOCAL AREA MODIFICATION	
				%	PRICE
DIVIDED LIGHT GRILLE	DIVIDED LIGHT GRILLES FOR WINDOWS **ADD**	SF	3.45		
SCREENS	FULL LENGTH INSECT SCREENS FOR WINDOWS **ADD**	SF	2.60		
ONE LIGHT PICTURE SASH & FRAME	• VINYL COVERED FRAME AND BRICK MOULDING • PRIMED SASH WITH 1/2" INSULATING GLASS • INTERIOR TRIM, STOPS, STOOL AND APRON				
	3-0 X 4-2	EA	477.00		
	4-6	EA	504.00		
	4-0 X 4-2	EA	531.00		
	4-6	EA	555.00		
	5-2	EA	609.00		
	5-4 X 4-6	EA	666.00		
PICTURE WINDOW WITH FLANKERS	PICTURE WINDOW ABOVE WITH DOUBLE HUNG FRAMES AND SASH, INCLUDING ALL TRIM, ETC. TOTAL THE COSTS FOR DOUBLE HUNG AND PICTURE WINDOW AND **ADD** EA = TOTAL UNIT	EA	18.00		

13

13. WOOD AWNING WINDOWS

SPECIFICATIONS		UNIT	PRICE	LOCAL AREA MODIFICATION	
				%	PRICE
AWNING WINDOW, SINGLE GLAZED	INCLUDING WEATHERSTRIPPING, EXTERIOR BRICK MOULD, FRAME, SASH, DRIP CAP, SCREEN, HARDWARE, INTERIOR TRIM				
	(1) 2-8 X 1-8	EA	267.00		
	(2) 3-4	EA	384.00		
	(3) 5-0	EA	501.00		
	(1) 2-4	EA	294.00		
	(2) 4-8	EA	417.00		
	(3) 7-0	EA	549.00		
	(1) 4-0 X 1-8	EA	333.00		
	(2) 3-4	EA	489.00		
	(3) 5-0	EA	624.00		
	(1) 2-4	EA	354.00		
	(2) 4-8	EA	501.00		
	(3) 7-0	EA	654.00		
AWNING WINDOW, INSULATED GLASS	INSULATED GLASS, INCLUDING WEATHERSTRIPPING, EXTERIOR BRICK MOULD, FRAME, SASH, DRIP CAP, SCREEN, HARDWARE, INTERIOR TRIM				
	(1) 2-8 X 1-8	EA	291.00		
	(2) 3-4	EA	420.00		
	(3) 5-0	EA	549.00		
	(1) 2-4	EA	321.00		
	(2) 4-8	EA	456.00		
	(3) 7-0	EA	600.00		
	(1) 4-0 X 1-8	EA	363.00		
	(2) 3-4	EA	516.00		
	(3) 5-0	EA	684.00		
	(1) 2-4	EA	384.00		
	(2) 4-8	EA	546.00		
	(3) 7-0	EA	720.00		

SPECIFICATIONS		UNIT	PRICE	LOCAL AREA MODIFICATION %	LOCAL AREA MODIFICATION PRICE
VENTING, SINGLE GLAZED	• STANDARD GLASS • EXTERIOR FRAME AND BRICK MOULDING • SASH • INTERIOR CASING • STOOL CAP AND APRON • ROTO OPERATOR • HARDWARE • SCREEN • ONE PANEL				
	Sash Size				
	1-11 X 2-6	EA	225.00		
	3-1	EA	243.00		
	3-6	EA	261.00		
	4-6	EA	276.00		
	5-6	EA	300.00		
	2-4 X 2-6	EA	243.00		
	3-1	EA	264.00		
	3-6	EA	267.00		
	4-6	EA	285.00		
VENTING, INSULATED GLASS, BUILDER QUALITY	SAME AS ABOVE WITH INSULATED GLASS, MEDIUM QUALITY				
	Sash Size				
	1-11 X 2-6	EA	273.00		
	3-1	EA	282.00		
	4-6	EA	336.00		
	5-6	EA	381.00		
	2-4 X 2-6	EA	294.00		
	3-1	EA	306.00		
	3-6	EA	324.00		
	4-6	EA	333.00		
	5-6	EA	387.00		
VENTING, INSULATED GLASS, PREMIUM QUALITY	• INSULATED GLASS • VINYL COVERED SASH AND FRAME • INTERIOR CASING • STOOL CAP AND APRON • ROTO OPERATOR • HARDWARE • SCREEN • ONE PANEL				
	Sash Size				
	1-11 X 2-6	EA	345.00		
	3-1	EA	357.00		
	4-6	EA	429.00		
	5-6	EA	486.00		
	2-4 X 2-6	EA	372.00		
	3-0	EA	387.00		
	4-6	EA	423.00		
	5-6	EA	495.00		
STATIONARY	SAME AS ABOVE WITHOUT ROTO OPERATOR **DEDUCT**	EA	10.50		

13. WOOD SLIDING WINDOWS

SPECIFICATIONS		UNIT	PRICE	LOCAL AREA MODIFICATION	
				%	PRICE
SINGLE GLAZED	• BUILDER QUALITY SASH AND FRAME • SINGLE GLAZED • SCREEN • APRON, STOOL AND TRIM • HARDWARE				
	3-0 X 3-0	EA	279.00		
	3-6	EA	288.00		
	4-0 X 3-6	EA	324.00		
	4-0	EA	330.00		
	5-0 X 3-6	EA	363.00		
	4-0	EA	372.00		
	5-0	EA	390.00		
	6-0 X 4-0	EA	492.00		
	5-0	EA	510.00		
INSULATED GLASS	• BUILDER QUALITY SASH AND FRAME • INSULATED GLASS • SCREEN • APRON, STOOL AND TRIM • HARDWARE				
	3-0 X 3-0	EA	366.00		
	3-6	EA	375.00		
	4-0 X 3-6	EA	429.00		
	4-0	EA	438.00		
	5-0 X 3-6	EA	483.00		
	4-0	EA	498.00		
	5-0	EA	522.00		
	6-0 X 4-0	EA	663.00		
	5-0	EA	687.00		
PREMIUM GRADE	• VINYL COVERED SASH AND FRAME • INSULATED GLASS • SCREEN • APRON, STOOL AND TRIM • HARDWARE				
	3-0 X 3-0	EA	474.00		
	3-6	EA	489.00		
	4-0 X 3-6	EA	558.00		
	4-0	EA	573.00		
	5-0 X 3-6	EA	633.00		
	4-0	EA	654.00		
	5-0	EA	684.00		
	6-0 X 4-0	EA	876.00		
	5-0	EA	909.00		

SPECIFICATIONS		UNIT	PRICE	LOCAL AREA MODIFICATION	
				%	PRICE
BUILDER GRADE	• PRIMED WOOD EXTERIOR SASH, FRAME AND TRIM • INSULATED GLASS • 2-1/4" INTERIOR ARCHED CASING • CURVED EXTENSION JAMBS • STOOL CAP				
	Width ***Height***				
	2-2 x 1-4	EA	375.00		
	2-6 x 1-4	EA	429.00		
	2-10 x 1-4	EA	450.00		
	3-2 x 1-4	EA	474.00		
	3-6 x 2-0	EA	498.00		
PREMIUM GRADE	• VINYL CLAD EXTERIOR SASH, FRAME AND TRIM • INSULATED GLASS • 2-1/4" INTERIOR ARCHED CASING • CURVED EXTENSION JAMBS • STOOL CAP				
	Width ***Height***				
	2-2 x 1-4	EA	498.00		
	2-6 x 1-4	EA	576.00		
	2-10 x 1-4	EA	609.00		
	3-2 x 1-4	EA	642.00		
	3-6 x 2-0	EA	687.00		
DIVIDED LIGHT GRILLES	DIVIDED LIGHT GRILLES FOR ABOVE UNITS				
	Width ***Height***				
	2-2 x 1-4	EA	27.00		
	2-6 x 1-4	EA	30.00		
	2-10 x 1-4	EA	33.00		
	3-2 x 1-4	EA	39.00		
	3-6 x 2-0	EA	42.00		

13. BAY WINDOW

SPECIFICATIONS		UNIT	PRICE	LOCAL AREA MODIFICATION	
				%	PRICE
BAY WINDOW, PREMIUM QUALITY	PRE-FABRICATED 30° BAY WINDOW IN EXISTING OPENING • TWO 1-8 X 4-6 WOOD DOUBLE HUNG, ONE STATIONARY 4-4 X 4-6 • HIGH PERFORMANCE INSULATED GLASS • VINYL CLAD • PROJECTION 14" • STOPS • INTERIOR CASINGS • EXTERIOR BRICK MOULDING	EA	2178.00		
	PRE-FABRICATED 45° BAY WINDOW IN EXISTING OPENING • TWO 1-8 X 4-6 WOOD DOUBLE HUNG, ONE STATIONARY 4-4 X 4-6 • HIGH PERFORMANCE INSULATED GLASS • VINYL CLAD • PROJECTION 20" • STOPS • INTERIOR CASINGS • EXTERIOR BRICK MOULDING	EA	2292.00		
	PRE-FABRICATED 90° BOX BAY WINDOW IN EXISTING OPENING • TWO 1-8 X 4-6 WOOD DOUBLE HUNG, ONE STATIONARY 4-4 X 4-6 • HIGH PERFORMANCE INSULATED GLASS • VINYL CLAD • PROJECTION 23" • STOPS • INTERIOR CASINGS • EXTERIOR BRICK MOULDING	EA	2328.00		
	BUILD ROOF OVER PROJECTION WITH 2 X 4 FRAMING, 1/2" PLYWOOD ROOF SHEATHING, 3/4" X 3/4" COVE FASCIA	EA	411.00		
	ON EXISTING SHEATHING, INSTALL FELT PAPER, FLASHING AND ASPHALT SHINGLES	EA	243.00		
	3/4" PLYWOOD HEADBOARD, TWO 3/4" PLYWOOD SEAT BOARDS AND TWO KNEE BRACES EA = ALL THREE ITEMS	EA	369.00		

13. BOW WINDOW, GREENHOUSE WINDOW

SPECIFICATIONS		UNIT	PRICE	LOCAL AREA MODIFICATION	
				%	PRICE
BOW WINDOW PREMIUM QUALITY	PRE-FABRICATED BOW WINDOW IN EXISTING OPENING • TWO 2-0 X 5-6 VENTING WOOD CASEMENT SASH AND TWO STATIONARY • HIGH PERFORMANCE INSULATED GLASS • REMOVABLE MUNTINS • ROLL-UP HARDWARE • PROJECTION 10" • INTERIOR CASINGS • EXTERIOR BRICK MOULDING • SCREEN	EA	2526.00		
	SAME AS ABOVE, ALUMINUM CLAD WINDOW	EA	2976.00		
	BUILD ROOF OVER PROJECTION WITH 2 X 4 FRAMING, 1/2" PLYWOOD ROOF SHEATHING, 3/4" X 3/4" COVE FASCIA	EA	258.00		
	ON EXISTING SHEATHING, INSTALL FELT PAPER, FLASHING AND ASPHALT SHINGLES	EA	234.00		
	3/4" PLYWOOD HEADBOARD, TWO 3/4" PLYWOOD SEAT BOARDS AND TWO KNEE BRACES EA = ALL THREE ITEMS	EA	369.00		
GREEN-HOUSE WINDOW	• BUMP-OUT WINDOW • 1/2" INSULATED GLASS • ADJUSTABLE SHELVES • ALUMINUM FRAME • PROJECTS OUT 14" • BRONZE OR WHITE				
	2-0 X 3-0	EA	585.00		
	4-0	EA	684.00		
	5-0	EA	786.00		
	3-0 X 3-0	EA	687.00		
	4-0	EA	798.00		
	5-0	EA	891.00		
	6-0	EA	999.00		
	4-0 X 3-0	EA	792.00		
	4-0	EA	897.00		
	5-0	EA	1011.00		
	6-0	EA	1107.00		

13

13. ROOF WINDOW

SPECIFICATIONS		UNIT	PRICE	LOCAL AREA MODIFICATION %	LOCAL AREA MODIFICATION PRICE
	NOTE: SEE PAGE 71 FOR ROOF WINDOW INSTALLATION IN EXISTING ROOF				
ROOF WINDOW	INSTALL DURING NEW CONSTRUCTION • CENTER PIVOT • ALUMINUM CLAD WOOD FRAME • INSULATED GLASS • EXTERIOR AWNING • INSECT SCREEN • BUILD CURB ON EXISTING FRAMED OPENING • INTERIOR WELL & TRIM				
	Outside Frame				
	30-3/4" X 38-1/2"	EA	954.00		
	30-3/4" X 55"	EA	1092.00		
	36-7/8" X 62-7/8"	EA	1299.00		
	44-7/8" X 46-1/2"	EA	1251.00		
	27-1/2" X 46-1/2"	EA	987.00		
	21-5/8" X 38-1/2"	EA	831.00		
	52-3/4" X 38-1/2"	EA	1296.00		
	52-3/4" X 55"	EA	1461.00		
	21-5/8" X 27-1/2"	EA	783.00		
	SAME AS ABOVE, TOP-HUNG				
	30-3/4" X 38-1/2"	EA	996.00		
	30-3/4" X 55"	EA	1137.00		
	44-7/8" X 46-1/2"	EA	1302.00		
	21-5/8" X 38-1/2"	EA	867.00		

SPECIFICATIONS		UNIT	PRICE	LOCAL AREA MODIFICATION %	LOCAL AREA MODIFICATION PRICE
	NOTE: SEE PAGE 70 FOR SKYLIGHT INSTALLATION IN EXISTING ROOF				
HATCH TYPE SKYLIGHT	• INSULATED GLASS • 1/8" TEMPERED OVER 1/4" LAMINATED GLASS WITH LOW E COATING • DOUBLE ALUMINUM WINDOW FRAME • COPPER FLASHING COVERS • MARINE GRADE PLYWOOD LINER BOX				
FIXED SKYLIGHT	***Rough Opening Sizes***				
	22" X 30"	EA	666.00		
	X 45-1/2"	EA	762.00		
	30" X 22"	EA	666.00		
	X 30"	EA	726.00		
	X 45-1/2"	EA	861.00		
	45-1/2" X 30"	EA	861.00		
	X 45-1/2"	EA	1017.00		
VENTILAT-ING SKYLIGHT W/INSECT SCREEN	***Rough Opening Sizes***				
	22" X 30"	EA	735.00		
	X 45-1/2"	EA	843.00		
	30" X 22"	EA	735.00		
	X 30"	EA	795.00		
	X 45-1/2"	EA	954.00		
	45-1/2" X 30"	EA	954.00		
	X 45-1/2"	EA	1119.00		

13. ALUMINUM SINGLE HUNG WINDOWS

SPECIFICATIONS			UNIT	PRICE	LOCAL AREA MODIFICATION	
					%	PRICE
ALUMINUM SINGLE HUNG, DOUBLE GLAZED	INCL. ALUMINUM FRAME, WEATHERSTRIPPING, HARDWARE AND SCREEN, DOUBLE GLAZED					
	2-0 X	3-0	EA	133.50		
		4-0	EA	139.50		
		5-0	EA	169.50		
		6-0	EA	178.50		
	2-8 X	3-0	EA	145.50		
		4-0	EA	156.00		
		5-0	EA	183.00		
		6-0	EA	199.50		
	3-0 X	3-0	EA	163.50		
		4-0	EA	180.00		
		5-0	EA	186.00		
		6-0	EA	205.50		
	4-0 X	3-0	EA	172.50		
		4-0	EA	187.50		
		5-0	EA	211.50		
		6-0	EA	246.00		
SINGLE GLAZED	SAME AS ABOVE, SINGLE GLAZED					
	2-0 X	2-4	EA	96.00		
		3-0	EA	102.00		
		5-0	EA	130.50		
	2-8 X	3-0	EA	115.50		
		4-0	EA	129.00		
		4-4	EA	130.50		
		5-0	EA	138.00		
	3-0 X	3-0	EA	120.00		
		4-0	EA	135.00		
		4-4	EA	136.50		
		6-0	EA	142.50		

13. ALUMINUM SLIDING AND AWNING WINDOWS

SPECIFICATIONS				UNIT	PRICE	LOCAL AREA MODIFICATION	
						%	PRICE
ALUMINUM SLIDING WINDOW, SINGLE GLAZED	INCLUDING ALUMINUM FRAME, WEATHERSTRIPPING, HARDWARE, SCREEN & INTERIOR TRIM, SINGLE GLAZED						
		2-0 X	3-0	EA	159.00		
		3-0 X	3-0	EA	168.00		
			4-0	EA	186.00		
		4-0 X	3-0	EA	177.00		
			3-6	EA	186.00		
			4-0	EA	195.00		
		5-0 X	3-6	EA	201.00		
			4-0	EA	216.00		
		6-0 X	4-0	EA	228.00		
ALUMINUM SLIDING WINDOW, INSULATED GLASS	INCLUDING ALUMINUM FRAME, WEATHERSTRIPPING, HARDWARE, SCREEN & INTERIOR TRIM, INSULATED GLASS						
		2-0 X	3-0	EA	189.00		
		3-0 X	3-0	EA	213.00		
			4-0	EA	246.00		
		4-0 X	3-0	EA	234.00		
			3-6	EA	255.00		
			4-0	EA	273.00		
		5-0 X	3-6	EA	291.00		
			4-0	EA	315.00		
		6-0 X	4-0	EA	348.00		
ALUMINUM AWNING	INCLUDING ALUMINUM FRAME, SCREEN AND INTERIOR TRIM						
	Panels						
	1	2-1 X	2-1	EA	177.00		
	2		3-1	EA	189.00		
	3		4-1	EA	228.00		
	4		5-1	EA	240.00		
	5		6-1	EA	261.00		
	1	3-1 X	2-1	EA	183.00		
	2		3-1	EA	210.00		
	3		4-1	EA	240.00		
	4		5-1	EA	276.00		
	5		6-1	EA	306.00		
	1	4-1 X	2-1	EA	210.00		
	2		3-1	EA	231.00		
	3		4-1	EA	276.00		
	4		5-1	EA	294.00		
	5		6-1	EA	357.00		

13. DOUBLE HUNG REPLACEMENT WINDOWS

SPECIFICATIONS		UNIT	PRICE	LOCAL AREA MODIFICATION %	LOCAL AREA MODIFICATION PRICE
	THE REPLACEMENT WINDOW COSTS SHOWN ON THIS AND THE FOLLOWING PAGES ARE FOR INSTALLATION IN OPENINGS APPROXIMATELY THE SAME SIZE AS THE REPLACEMENT WINDOWS				
WINDOW OPENING SIZE ADJUST-MENT	FOR OPENINGS UP TO 2-1/2" LARGER IN HEIGHT OR WIDTH THAN REPLACEMENT WINDOW, FRAME OPENING TO WINDOW SIZE WITH FILLER STRIPS AND APPLY WIDER CASINGS AND TRIM **ADD**	EA	48.00		
	TO REDUCE OPENING SIZE IN STUCCO WALL, APPLY STUCCO ON EXTERIOR, AND DRYWALL OR TRIM ON INTERIOR **ADD**	EA	99.00		
WOOD DOUBLE HUNG	REPLACE EXISTING WOOD DOUBLE HUNG WIN-DOW WITH NEW WOOD DOUBLE HUNG WINDOW IN WOOD OR BRICK VENEER EXTERIOR WALL • REMOVE INTERIOR CASING AND EXTERIOR WINDOW TRIM • REMOVE EXISTING WINDOW AND FRAME FROM WALL WITH DRYWALL OR PLASTER INTERIOR • INSTALL NEW WOOD DOUBLE HUNG TWO-LIGHT WINDOW AND FRAME WITH INSULATED GLASS AND SCREEN • REPLACE INTERIOR AND EXTERIOR TRIM AND CASING • TOUCH UP PAINT TO MATCH EXISTING				
	UP TO 72 UI	EA	381.00		
	73 TO 82 UI	EA	450.00		
	83 TO 92 UI	EA	516.00		
	93 TO 101 UI	EA	582.00		
	UI = ADD TOGETHER THE WIDTH IN INCHES AND THE HEIGHT IN INCHES				
VINYL DOUBLE HUNG WINDOW	SAME AS ABOVE EXCEPT REPLACEMENT WINDOW IS VINYL WITH INSULATED GLASS AND SCREEN				
	UP TO 72 UI	EA	327.00		
	73 TO 82 UI	EA	372.00		
	83 TO 92 UI	EA	415.00		
	93 TO 101 UI	EA	453.00		
WINDOWS IN MASONRY WALL	ANY OF THE ABOVE WINDOWS INSTALLED IN BRICK OR BRICK & BLOCK WALL **ADD**	EA	24.00		
DIVIDED LIGHT GRILLE	DIVIDED LIGHT GRILLE FOR ABOVE WINDOWS **ADD**	SF	3.45		
	SF= OVERALL SIZE OF WINDOW SASH				

13

13. WOOD SASH & CASEMENT REPLACEMENT WINDOWS

SPECIFICATIONS		UNIT	PRICE	LOCAL AREA MODIFICATION %	LOCAL AREA MODIFICATION PRICE
WOOD CASEMENT WINDOW	REPLACE EXISTING WOOD DOUBLE HUNG OR CASEMENT WINDOW WITH NEW WOOD CASEMENT WINDOW IN WOOD OR BRICK VENEER EXTERIOR WALL • REMOVE INTERIOR CASING AND EXTERIOR WINDOW TRIM • REMOVE EXISTING WINDOW AND FRAME FROM WALL • INSTALL NEW WOOD CASEMENT WINDOW WITH INSULATED GLASS AND SCREEN • REPLACE INTERIOR AND EXTERIOR TRIM AND CASING • TOUCH UP PAINT TO MATCH EXISTING				
	UP TO 72 UI	EA	393.00		
	73 TO 82 UI	EA	463.50		
	83 TO 92 UI	EA	529.50		
	93 TO 101 UI	EA	597.00		
	UI = ADD TOGETHER THE WIDTH IN INCHES AND THE HEIGHT IN INCHES				
VINYL CASEMENT WINDOW	SAME AS ABOVE EXCEPT REPLACEMENT WINDOW IS VINYL WITH INSULATED GLASS AND SCREEN				
	UP TO 72 UI	EA	321.00		
	73 TO 82 UI	EA	361.50		
	83 TO 92 UI	EA	408.00		
	93 TO 101 UI	EA	462.00		
WINDOWS IN MASONRY WALL	ANY OF THE ABOVE WINDOWS INSTALLED IN BRICK OR BRICK & BLOCK WALL **ADD**	EA	24.00		
DIVIDED LIGHT GRILLE	DIVIDED LIGHT GRILLE FOR ABOVE WINDOWS **ADD** SF= OVERALL SIZE OF WINDOW SASH	SF	3.45		
REMOVE STEEL CASEMENT AND REPLACE	REMOVE EXISTING STEEL CASEMENT WINDOW SET IN MASONRY WALL AND REPLACE WITH ANY OF ABOVE WINDOWS **ADD**	EA	15.00		
WOOD SASH REPLACEMENT ONLY	DOUBLE HUNG REPLACEMENT SASH AND JAMB LINERS • 1/2" INSULATING GLASS • 2 LIGHTS PER OPENING				
	UP TO 72 UI	EA	208.50		
	73 TO 82 UI	EA	243.00		
	83 TO 92 UI	EA	282.00		
	93 TO 101 UI	EA	315.00		

13. VINYL SLIDER & BASEMENT REPLACEMENT WINDOWS

SPECIFICATIONS		UNIT	PRICE	LOCAL AREA MODIFICATION	
				%	PRICE
VINYL SLIDER	REPLACE EXISTING WOOD OR METAL WINDOW WITH VINYL BYPASS SLIDER UNIT • REMOVE INTERIOR CASING AND EXTERIOR WINDOW TRIM • REMOVE EXISTING WINDOW AND FRAME FROM WOOD OR BRICK VENEER EXTERIOR WALL • INSTALL NEW VINYL BYPASS SLIDER UNIT WITH INSULATED GLASS AND SCREEN • REPLACE INTERIOR AND EXTERIOR TRIM AND CASING • TOUCH UP PAINT TO MATCH EXISTING				
	UP TO 72 UI	EA	351.00		
	73 TO 82 UI	EA	394.50		
	83 TO 92 UI	EA	438.00		
	93 TO 101 UI	EA	469.50		
	102 TO 112 UI	EA	513.00		
	UI = ADD TOGETHER THE WIDTH IN INCHES AND THE HEIGHT IN INCHES				
WINDOWS IN MASONRY WALL	ANY OF THE ABOVE WINDOWS INSTALLED IN BRICK OR BRICK & BLOCK WALL **ADD**	EA	24.00		
BASEMENT WINDOW	REPLACE EXISTING AWNING STYLE METAL BASEMENT WINDOW WITH AWNING STYLE VINYL BASEMENT WINDOW • REMOVE EXISTING WINDOW FROM MASONRY WALL • INSTALL NEW AWNING STYLE VINYL BASEMENT WINDOW WITH INSULATED GLASS AND SCREEN				
	UP TO 72 UI	EA	261.00		

13

SPECIFICATIONS		UNIT	PRICE	LOCAL AREA MODIFICATION %	LOCAL AREA MODIFICATION PRICE
ALUMINUM COMBINATION WINDOWS	TRIPLE TRACK, SELF STORING				
	2 GLASS INSERTS AND 1 SCREEN INSERT				
	WEATHERSTRIPPING				
	MILL FINISH				
	UP TO 101 UNITED INCHES (ADD HEIGHT OF WINDOW PLUS WIDTH OF WINDOW)				
	ECONOMY	EA	105.00		
	MEDIUM	EA	114.00		
	PREMIUM	EA	135.00		
	FOR WINDOWS OVER 101 UNITED INCHES **ADD**	PER INCH	1.05		
	WHITE FINISH **ADD**	EA	3.00		
	COLOR FINISH **ADD**	EA	4.50		
CASEMENT	16-3/4" X 23-5/8"	EA	75.00		
	36-1/8"	EA	78.00		
	48-3/8"	EA	87.00		

14. PLUMBING

SPECIFICATIONS		UNIT	PRICE	LOCAL AREA MODIFICATION %	LOCAL AREA MODIFICATION PRICE
	THE PRICES SHOWN IN THIS SECTION INCLUDE A PLUMBING SUBCONTRACTOR'S OVERHEAD & PROFIT				
HOSE BIBB	NEW OUTSIDE FROST-FREE SPIGOT (FAUCET)	EA	165.00		
	RE-LOCATE EXISTING HOSE BIBB UP TO 20 FEET	EA	150.00		
CUTOFF	INSTALL CUTOFF VALVES ON SINK LINES	EA	42.00		
GAS LINE	EXTEND GAS LINE INSIDE BUILDING WITHOUT BREAKING WALLS	LF	21.00		
REPLACE WATER PIPES	REPLACE HORIZONTAL GALVANIZED WATER PIPES IN BASEMENT WITH 3/4" COPPER PIPES, OPEN CEILING	EA PLUS	954.00		
	FULL BASEMENT	SF	.68		
	CRAWL SPACE	SF	.98		
	SF = FLOOR AREA				
	REPLACE VERTICAL GALVANIZED WATER PIPES TO EACH BATH OR KITCHEN WITH 3/4" COPPER, OPEN FLOORS AND WALLS				
	BASEMENT TO FIRST FLOOR	EA	597.00		
	BASEM'T TO SECOND FLOOR	EA	819.00		
VENT	RAISE EXISTING VENT THROUGH ROOF UP TO 10 FEET	EA	168.00		
	RUN NEW VENT THROUGH EXISTING ROOF	LF	21.00		
STACK	NEW STACK RUNNING FROM BASEMENT OR CRAWL SPACE UP AND VENTING THROUGH ROOF				
	ONE STORY	EA	459.00		
	TWO STORY	EA	654.00		
WASTE PIPE	EXTEND WASTE LINE FROM NEW STACK IN BASEMENT OR CRAWL SPACE				
	HANGING WASTE LINE	LF	51.00		
	BELOW GROUND WASTE LINE, INCLUDING EXCAVATION AND BACKFILL	LF	57.00		
	BELOW GROUND WASTE LINE, INCLUDING BREAKING CONCRETE FLOOR AND REPLACING	LF	60.00		
SEWER EJECTION PUMP	STANDARD SEWER EJECTION PUMP FOR BASEMENT BATH, WITH ADEQUATE ACCESS	EA	1164.00		
GRAY BOX	INSTALL WALL MOUNTED SUPPLY AND WASTE "GRAY BOX" IN LAUNDRY AREA	EA	351.00		

14

SPECIFICATIONS		UNIT	PRICE	LOCAL AREA MODIFICATION %	LOCAL AREA MODIFICATION PRICE
	NOTE: THE FOLLOWING INSTALLATIONS ARE FOR NEW WORK LOCATED WITHIN 5 FEET OF EXISTING STACK AND INCLUDE COST OF ROUGH-IN, FIXTURES AND FITTINGS, AND COMPLETE INSTALLATION.				
OVER 5 FEET FROM STACK	IF EXISTING STACK IS OVER 5 FEET FROM THE NEW W.C. INSTALLATION, A NEW STACK WILL PROBABLY BE REQUIRED, RUNNING FROM BASEMENT UP & THROUGH EXISTING ROOF. SEE PAGE 202.				
INSTALL TWO FIXTURES	ANY TWO FIXTURES ABOVE, ADD TOTAL COST OF BOTH AND **DEDUCT**	SET	10%		
INSTALL THREE OR MORE FIXTURES	THREE OR MORE OF ABOVE FIXTURES, ADD TOTAL COST AND **DEDUCT**	SET	15%		
WATER CLOSET	2-PIECE FLOOR MOUNTED VITREOUS CHINA WATER CLOSET WITHIN 5 FEET OF EXISTING STACK				
	WHITE	EA	1299.00		
	COLOR	EA	1359.00		
	1-PIECE FLOOR MOUNTED VITREOUS CHINA WATER CLOSET WITHIN 5 FEET OF EXISTING STACK				
	WHITE	EA	1764.00		
	COLOR	EA	1929.00		

14

14. BATHTUB, SHOWER, BIDET — NEW WORK

SPECIFICATIONS		UNIT	PRICE	LOCAL AREA MODIFICATION %	LOCAL AREA MODIFICATION PRICE
	NOTE: THE FOLLOWING INSTALLATIONS ARE FOR NEW WORK LOCATED WITHIN 5 FEET OF EXISTING STACK AND INCLUDE COST OF ROUGH-IN, FIXTURES AND FITTINGS, AND COMPLETE INSTALLATION.				
OVER 5 FEET FROM STACK	GOING AWAY FROM STACK OVER 5 FEET IN INSTALLATIONS ON THIS AND THE PREVIOUS PAGE **ADD**	LF	54.00		
	LF = TOTAL DISTANCE FROM EXISTING STACK				
BATHTUB	5-FOOT BATHTUB, FAUCET @ $60, DRAIN ASSEMBLY @ $75				
	CAST IRON @ $250	EA	1689.00		
	STEEL @ $150	EA	1479.00		
	FIBERGLASS @ $170	EA	1539.00		
	ENAMELED CAST IRON 5-FOOT BATHTUB WITH BUILT-IN WHIRLPOOL, INCLUDING ELECTRICAL CONNECTION	EA	4569.00		
FIBERGLASS WHIRLPOOL	FIBERGLASS WHIRLPOOL TUB, 5-FOOT X 32" X 18" DEEP, 3 JETS, INCLUDING SKIRT, INCLUDING ELECTRICAL CONNECTION	EA	4275.00		
STALL SHOWER	ROUGH AND INSTALL PLUMBING FOR STALL SHOWER, INCLUDING VINYL OR RUBBER PAN, READY FOR WALL AND FLOOR TILE, FAUCET @ $45, DRAIN ASSEMBLY @ $65	EA	1065.00		
	ROUGH AND INSTALL FIBERGLASS SHOWER STALL @ $300, FAUCET @ $45, DRAIN ASSEMBLY @ $65	EA	1659.00		
	ROUGH AND INSTALL FIBERGLASS TUB/SHOWER UNIT @ $337, FAUCET @ $55, DRAIN ASSEMBLY @ $65	EA	1749.00		
BIDET	INSTALL FLOOR MOUNTED BIDET, FITTINGS				
	WHITE	EA	2034.00		
	COLOR	EA	2229.00		

14

SPECIFICATIONS		UNIT	PRICE	LOCAL AREA MODIFICATION %	LOCAL AREA MODIFICATION PRICE
LAVATORY, WALL HUNG	WALL HUNG LAVATORY, PORCELAIN ENAMEL ON CAST IRON, 20" X 18"				
	WHITE	EA	1059.00		
	COLOR	EA	1095.00		
	VITREOUS CHINA, 19" X 17"				
	WHITE	EA	999.00		
	COLOR	EA	1044.00		
PEDESTAL SINK	PEDESTAL MOUNTED, VITREOUS CHINA, 27" X 23"				
	WHITE	EA	1335.00		
	COLOR	EA	1395.00		
LAVATORY SET IN VANITY BASE	PORCELAIN ENAMEL ON CAST IRON, 20" X 10"				
	WHITE	EA	1119.00		
	COLOR	EA	1164.00		
	ENAMELED STEEL, 10" X 17"				
	WHITE	EA	999.00		
	COLOR	EA	1044.00		
	VITREOUS CHINA, 20" X 16"				
	WHITE	EA	1050.00		
	COLOR	EA	1095.00		
VANITY BASE	PRE-FINISHED VANITY BASE, BUILDER QUALITY, 21" DEEP				
	Doors ***Drawers*** ***Width***				
	2 0 24"	EA	390.00		
	2 0 30"	EA	471.00		
	2 0 36"	EA	609.00		
	3 3 42"	EA	690.00		
	3 3 48"	EA	789.00		
	3 3 60"	EA	1014.00		
VANITY TOPS	CULTURED MARBLE TOP AND 4" SPLASH	LF	51.00		

14. REPLACE BATHROOM FIXTURES

SPECIFICATIONS		UNIT	PRICE	LOCAL AREA MODIFICATION	
				%	PRICE
	NOTE: REMOVE EXISTING BATHROOM FIXTURES AND REPLACE WITH NEW FIXTURES IN SAME LOCATION				
REPLACE WATER CLOSET	2-PIECE FLOOR MOUNTED VITREOUS CHINA WATER CLOSET				
	WHITE	EA	384.00		
	COLOR	EA	459.00		
	1-PIECE FLOOR MOUNTED VITREOUS CHINA WATER CLOSET				
	WHITE	EA	834.00		
	COLOR	EA	1023.00		
REPLACE BIDET	FLOOR MOUNTED BIDET, BRASS FITTINGS				
	WHITE	EA	1455.00		
	COLOR	EA	1584.00		
REPLACE BATHTUB	5-FOOT BATHTUB, INCL. FAUCET @ $60, DRAIN ASSEMBLY @ $75				
	CAST IRON @ $250	EA	1290.00		
	STEEL @ $150	EA	984.00		
	FIBERGLASS @ $170	EA	1065.00		
	ENAMELED CAST IRON 5-FOOT BATHTUB WITH BUILT-IN WHIRLPOOL, INCLUDING ELECTRICAL CONNECTION	EA	3579.00		
	FIBERGLASS WHIRLPOOL TUB, 5-FOOT X 32" X 18" DEEP, 3 JETS, INCLUDING SKIRT, INCLUDING ELECTRICAL CONNECTION	EA	3234.00		
STALL SHOWER	RE-ROUGH AND INSTALL PLUMBING FOR STALL SHOWER, INCLUDING VINYL OR RUBBER PAN, READY FOR WALL AND FLOOR TILE, FAUCET @ $45, DRAIN ASSEMBLY @ $65	EA	729.00		
REPLACE SHOWER PAN	REMOVE EXISTING DEFECTIVE SHOWER PAN (AFTER WALL AND FLOOR TILE HAVE BEEN REMOVED BY OTHERS), AND INSTALL NEW SHOWER PAN WITH DRAIN ASSEMBLY AND DRAIN COVER FOR COMPLETE JOB REPLACEMENT OF SHOWER PAN, SEE PAGE 64.	EA	720.00		

14

14. REPLACE BATHROOM FIXTURES

SPECIFICATIONS		UNIT	PRICE	LOCAL AREA MODIFICATION %	LOCAL AREA MODIFICATION PRICE
REPLACE LAVATORY	WALL HUNG LAVATORY, PORCELAIN ENAMEL ON CAST IRON, 20" X 18"				
	WHITE	EA	510.00		
	COLOR	EA	555.00		
	VITREOUS CHINA, 19" X 17"				
	WHITE	EA	465.00		
	COLOR	EA	510.00		
	PEDESTAL MOUNTED, VITREOUS CHINA, 27" X 23"				
	WHITE	EA	735.00		
	COLOR	EA	780.00		
REPLACE LAVATORY SET IN VANITY BASE	PORCELAIN ENAMEL ON CAST IRON, 20" X 10"				
	WHITE	EA	456.00		
	COLOR	EA	501.00		
	ENAMELED STEEL, 10" X 17"				
	WHITE	EA	387.00		
	COLOR	EA	432.00		
	VITREOUS CHINA, 20" X 16"				
	WHITE	EA	420.00		
	COLOR	EA	465.00		
REPLACE VANITY BASE	PRE-FINISHED VANITY BASE, BUILDER QUALITY, 21" DEEP				
	Doors ***Drawers*** ***Width***				
	2 0 24"	EA	420.00		
	2 0 30"	EA	504.00		
	2 0 36"	EA	639.00		
	3 3 42"	EA	720.00		
	3 3 48"	EA	822.00		
	3 3 60"	EA	1044.00		
REPLACE VANITY TOP	CULTURED MARBLE TOP AND 4" SPLASH	LF	51.00		
REMOVE & REPLACE FIXTURE	DISCONNECT AND REMOVE BATHROOM FIXTURE AND REPLACE IN SAME LOCATION AT LATER TIME				
	WATER CLOSET	EA	180.00		
	BIDET	EA	510.00		
	WALL HUNG OR PEDESTAL LAVATORY	EA	354.00		
	LAVATORY & VANITY	EA	249.00		

14

14. KITCHEN PLUMBING & APPLIANCES — NEW WORK

SPECIFICATIONS		UNIT	PRICE	LOCAL AREA MODIFICATION %	LOCAL AREA MODIFICATION PRICE
	THE FOLLOWING PRICES INCLUDE FIXTURES AND ALL MATERIALS AND LABOR FOR ROUGHING IN AND INSTALLATION. ELECTRICAL COSTS ARE **NOT** INCLUDED.				
KITCHEN SINK	ROUGH AND INSTALL KITCHEN SINK WITHIN 5 FEET FROM EXISTING STACK, INCLUDING FAUCET, SPRAY AND TWO STRAINERS	EA	1245.00		
BAR SINK	15" X 15" STAINLESS STEEL BAR SINK WITH FAUCET AND STRAINER, WITHIN 5 FEET OF EXISTING STACK	EA	639.00		
DISPOSER	1/2 HP DISPOSER INSTALLED AT EXISTING SINK LOCATION	EA	294.00		
DISH-WASHER	INSTALL DISHWASHER NEXT TO EXISTING SINK				
	ECONOMY	EA	780.00		
	BUILDER	EA	891.00		
	PREMIUM	EA	1026.00		
DROP WASTE	DROP WASTE FOR ABOVE DISPOSER OR DISH-WASHER	EA	195.00		
LAUNDRY TUB	POLYPROPYLENE 20 GAL. SINK, INCLUDING RUB-BER STOPPER, RAISED SOAP DISH AND DRAIN ASSEMBLY WITH FAUCET	EA	924.00		
CLOTHES WASHER	INSTALL CLOTHES WASHER				
	ECONOMY	EA	939.00		
	BUILDER	EA	1071.00		
	PREMIUM	EA	1191.00		
GAS DRYER	INSTALL GAS DRYER				
	ECONOMY	EA	822.00		
	BUILDER	EA	1002.00		
	PREMIUM	EA	1074.00		
GAS RANGE	GAS COOKTOP 28" X 20"	EA	513.00		
	35" X 20"	EA	732.00		
GAS OVEN	GAS WALL OVEN, SINGLE	EA	672.00		
	DOUBLE	EA	765.00		
	FREE STANDING GAS RANGE AND OVEN				
	ECONOMY	EA	813.00		
	BUILDER	EA	987.00		
	PREMIUM	EA	1410.00		
INSTANT HOT	INSTALL INSTANT HOT NEXT TO SINK	EA	474.00		
ICEMAKER	RUN WATER LINE FOR REFRIGERATOR ICEMAKER WITHIN 5 FEET OF SINK	EA	150.00		
	SAME AS ABOVE, MORE THAN 5 FEET FROM SINK **ADD** LF = TOTAL DISTANCE FROM EXISTING SINK	LF	2.40		

14

14. REPLACE KITCHEN FIXTURES & APPLIANCES

SPECIFICATIONS		UNIT	PRICE	LOCAL AREA MODIFICATION %	LOCAL AREA MODIFICATION PRICE
	THE FOLLOWING PRICES INCLUDE FIXTURES AND ALL MATERIALS AND LABOR FOR ROUGHING IN AND INSTALLATION. ELECTRICAL COSTS ARE **NOT** INCLUDED.				
	IF MORE THAN ONE FIXTURE OR APPLIANCE BELOW IS REMOVED AND REPLACED AT THE SAME TIME **DEDUCT** EA = EACH FIXTURE	EA	5%		
REPLACE KITCHEN SINK	REPLACE KITCHEN SINK IN SAME LOCATION	EA	675.00		
RELOCATE SINK	RE-LOCATE EXISTING KITCHEN SINK ON ANOTHER WALL IN KITCHEN, RE-ROUGH DRAIN AND VENT	EA	954.00		
REPLACE BAR SINK	REPLACE BAR SINK WITHIN 5 FEET OF EXISTING LINE	EA	480.00		
REPLACE DISPOSER	REPLACE DISPOSER WITH NEW 1/2 HP DISPOSER	EA	279.00		
REPLACE DISHWASHER	REPLACE DISHWASHER				
	ECONOMY	EA	639.00		
	BUILDER	EA	750.00		
	PREMIUM	EA	885.00		
REPLACE LAUNDRY TUB	REPLACE LAUNDRY TUB WITH NEW POLYPROPY-LENE 20 GAL. SINK	EA	354.00		
REPLACE WASHER	REPLACE CLOTHES WASHER				
	ECONOMY	EA	654.00		
	BUILDER	EA	798.00		
	PREMIUM	EA	900.00		
REPLACE GAS DRYER	REPLACE GAS DRYER				
	ECONOMY	EA	624.00		
	BUILDER	EA	798.00		
	PREMIUM	EA	879.00		
REPLACE GAS APPLIANCES	REPLACE GAS COOKTOP				
	28" X 20"	EA	210.00		
	35" X 20"	EA	447.00		
	REPLACE GAS WALL OVEN				
	SINGLE	EA	390.00		
	DOUBLE	EA	465.00		
	REPLACE FREE STANDING GAS RANGE AND OVEN				
	ECONOMY	EA	555.00		
	BUILDER	EA	720.00		
	PREMIUM	EA	1140.00		
REMOVE & REPLACE SAME FIXTURES	DISCONNECT AND REMOVE KITCHEN FIXTURES AND APPLIANCES AND REPLACE SAME FIXTURES IN SAME LOCATION AT LATER TIME SINK	EA	249.00		
	DISPOSER	EA	102.00		
	DISHWASHER	EA	204.00		
	GAS RANGE	EA	135.00		
	REFRIGERATOR W/ICE MAKER	EA	99.00		
	LAUNDRY TUB	EA	189.00		

14. HOT WATER HEATER

SPECIFICATIONS		UNIT	PRICE	LOCAL AREA MODIFICATION	
				%	PRICE
HOT WATER HEATER, NEW WORK	GAS HOT WATER HEATER, INCLUDING 2" FOAM INSULATION, FLUE AND ALL LABOR AND MATERIALS TO CONNECT WITH WATER AND GAS SUPPLY				
	30 GAL	EA	570.00		
	40 GAL	EA	642.00		
	50 GAL	EA	753.00		
	ELECTRIC WATER HEATER, INCLUDING ALL LABOR AND MATERIALS TO CONNECT TO WATER SUPPLY AND 220 VOLT LINE FROM ELECTRICAL PANEL BOX				
	30 GAL	EA	513.00		
	40 GAL	EA	585.00		
	52 GAL	EA	695.00		
	82 GAL	EA	1026.00		
REPLACE-MENT	REPLACE EXISTING GAS WATER HEATER WITH SAME SIZE NEW HEATER, **NO** PLUMBING MODIFICATIONS INCLUDED				
	30 GAL	EA	423.00		
	40 GAL	EA	495.00		
	50 GAL	EA	621.00		
	REPLACE EXISTING ELECTRIC WATER HEATER WITH SAME SIZE NEW HEATER, **NO** PLUMBING MODIFICATIONS INCLUDED				
	30 GAL	EA	405.00		
	40 GAL	EA	480.00		
	52 GAL	EA	602.00		
	82 GAL	EA	936.00		
REMOVE & REPLACE WATER HEATER	DISCONNECT AND REMOVE WATER HEATER AND REPLACE SAME HEATER IN SAME LOCATION AT A LATER TIME	EA	180.00		

14

SPECIFICATIONS		UNIT	PRICE	LOCAL AREA MODIFICATION %	LOCAL AREA MODIFICATION PRICE
	THE HEATING AND AIR CONDITIONING PRICES SHOWN IN THIS SECTION INCLUDE A SUBCONTRACTOR'S OVERHEAD AND PROFIT				
CONVECTOR	CONVECTORS ON EXISTING SYSTEM				
	FIRST FLOOR	EA	669.00		
	SECOND FLOOR	EA	906.00		
CAST IRON	NEW CAST IRON RADIATORS ON EXISTING SYSTEM				
	FIRST FLOOR	EA	864.00		
	SECOND FLOOR	EA	1104.00		
	USED RADIATOR ON EXISTING SYSTEM				
	FIRST FLOOR	EA	600.00		
	SECOND FLOOR	EA	780.00		
	REMOVE RADIATOR, RUN CAST IRON BASEBOARD HEAT ON SAME WALL, 8-FOOT STRETCH	EA	894.00		
	RE-LOCATE RADIATOR WITHIN 10 FEET OF SAME LOCATION, FLOOR, WALL AND CEILING PATCHING **NOT** INCLUDED	EA	510.00		
	EXTEND HEATING LINES TO NEW ADDITION, INSTALL ONE RADIATOR	EA	984.00		
BASEBOARD	ON EXISTING SYSTEM, 8-FOOT STRETCH				
	COPPER	EA	540.00		
	CAST IRON	EA	855.00		
	SAME AS ABOVE, 16-FOOT STRETCH				
	COPPER	EA	864.00		
	CAST IRON	EA	1155.00		

15. HOT WATER HEAT, OIL FIRED HEAT

SPECIFICATIONS		UNIT	PRICE	LOCAL AREA MODIFICATION	
				%	PRICE
CIRCULATING PUMP	INSTALL CIRCULATING PUMP ON EXISTING SYSTEM	EA	504.00		
SEPARATE ZONE	INSTALL SEPARATE ZONE AND THERMOSTAT, INCLUDING BY-PASS VALVE AND CIRCULATING PUMP	EA	891.00		
HEAT RISERS	EXTEND HEAT RISERS FROM FIRST TO SECOND FLOOR	EA	435.00		
DRAIN SYSTEM	DRAIN SYSTEM, DISCONNECT AND CAP OFF RADIATOR, FILL SYSTEM	EA	192.00		
BOILER	RE-BUILD CHAMBER IN EXISTING BOILER				
	MINIMUM	EA	414.00		
	MAXIMUM	EA	1125.00		
	NEW BOILER, GAS FIRED				
	100,000 BTU	EA	3420.00		
	125,000 BTU	EA	3915.00		
OIL BURNER	NEW OIL BURNER ON EXISTING BOILER	EA	858.00		
OIL TANK	INDOOR, 275 GALLONS	EA	1224.00		
	OUTDOOR, BURIED FIBERGLASS TANK AND PIPING, 550 GALLONS	EA	2172.00		
FURNACE	OIL FURNACE ON EXISTING DUCT SYSTEM AND FLUE, **NO** ELECTRICAL HOOKUP OR TANK INCLUDED				
	85,000 BTU	EA	3150.00		
	100,000 BTU	EA	3480.00		
	125,000 BTU	EA	3960.00		
	NEW BOILER, OIL FIRED				
	100,000 BTU	EA	3690.00		
	125,000 BTU	EA	4170.00		

SPECIFICATIONS			UNIT	PRICE	LOCAL AREA MODIFICATION	
					%	PRICE
ELECTRIC FURNACE	ELECTRIC FURNACE ON EXISTING DUCT SYSTEM, INCLUDING ALL LABOR AND MATERIALS, ON EXISTING CIRCUIT					
	10 KW		EA	1110.00		
	15 KW		EA	1500.00		
	20 KW		EA	1845.00		
	25 KW		EA	2610.00		
	30 KW		EA	2895.00		
ELECTRIC HEAT PUMP	ELECTRIC HEAT PUMP FOR HEAT AND AIR CONDITIONING, ON EXISTING SYSTEM. DETERMINE SIZE PER A/C LOAD, ON EXISTING CIRCUIT					
	10 KW, 2 TON		EA	6150.00		
	15 KW, 2.5 TON		EA	6885.00		
	19 KW, 3 TON		EA	7965.00		
	21 KW, 3.5 TON		EA	9420.00		
	28 KW, 4 TON		EA	9885.00		
	28 KW, 5 TON		EA	11,964.		
GAS FURNACE	GAS FURNACE, USING EXISTING DUCT SYSTEM & FLUE, INCLUDING GAS PIPE HOOKUP IF GAS LINE AT FURNACE, ON EXISTING CIRCUIT					
		Approximate Heating Area				
	60,000 BTU	750 SF	EA	2364.00		
	80,000 BTU	1000 SF	EA	2589.00		
	100,000 BTU	1200 SF	EA	2700.00		
	125,000 BTU	1350 SF	EA	2910.00		
	150,000 BTU	1500 SF	EA	3285.00		
	175,000 BTU	2000 SF	EA	3459.00		
	200,000 BTU	2500 SF	EA	4170.00		
	GAS WALL FURNACE, RECESSED		EA	1284.00		

15. ELECTRIC HEAT

SPECIFICATIONS		UNIT	PRICE	LOCAL AREA MODIFICATION	
				%	PRICE
BASEBOARD ELECTRIC, OPEN WALLS	INCLUDING BUILT-IN THERMOSTAT ON EXISTING SYSTEM WITHIN 10 LF OF BOX				
	4'-0" (1000 WATT)	EA	270.00		
	6'-0" (1500 WATT)	EA	309.00		
	8'-0" (2000 WATT)	EA	330.00		
	10'-0" (2500 WATT)	EA	369.00		
	SAME AS ABOVE, FISHED				
	4'-0" (1000 WATT)	EA	297.00		
	6'-0" (1500 WATT)	EA	324.00		
	8'-0" (2000 WATT)	EA	369.00		
	10'-0" (2500 WATT)	EA	402.00		
	OVER 10 LF FROM BOX				
	4'-0" (1000 WATT)	EA	327.00		
	6'-0" (1500 WATT)	EA	360.00		
	8'-0" (2000 WATT)	EA	408.00		
	10'-0" (2500 WATT)	EA	447.00		
WALL HEATER	WALL HEATER WITH BUILT-IN THERMOSTAT AND OUTLET, 12" X 12", 1250 WATTS	EA	285.00		
15 CEILING HEATER	CEILING HEATER WITH REMOTE THERMOSTAT AND SWITCH, 1250 WATTS	EA	354.00		
INFRA-RED HEATER	INFRA-RED HEATER IN CEILING ON SEPARATE SWITCH	EA	345.00		

SPECIFICATIONS				UNIT	PRICE	LOCAL AREA MODIFICATION %	LOCAL AREA MODIFICATION PRICE
AIR CONDITIONER	INSTALL SEPARATE A/C SYSTEM EACH FLOOR WITH AIR HANDLER INSIDE AND COMPRESSOR OUTSIDE		***Approximate Cooling Area***				
		2 TON	750 SF	EA	3750.00		
		2.5 TON	1500 SF	EA	4170.00		
		3 TON	2500 SF	EA	4725.00		
	INSTALL A/C SYSTEM ON EXISTING FURNACE, COMPRESSOR ON GROUND		***Approximate Cooling Area***				
		2 TON	750 SF	EA	2580.00		
		2.5 TON	1500 SF	EA	3009.00		
		3 TON	2500 SF	EA	3285.00		
		3.5 TON	3000 SF	EA	3645.00		
		4 TON	3500 SF	EA	3915.00		
		5 TON	4500 SF	EA	4845.00		
	FOR COMPRESSOR ON ROOF (INCLUDING CRANE COST) **ADD**			EA	1350.00		
	INCREASE BLOWER MOTOR CAPACITY TO ADD A/C			EA	195.00		
	CONDENSATE PUMP TO REMOVE EVAPORATOR WATER FROM A/C			EA	195.00		
WINDOW UNITS	HIGH EFFICIENCY AIR CONDITIONING AND COOLING UNITS ON EXISTING SERVICE		***Approximate Cooling Area***				
		5,600 BTU	190 SF	EA	720.00		
		9,000 BTU	420 SF	EA	855.00		
		13,000 BTU	720 SF	EA	1080.00		
		15,000 BTU	875 SF	EA	1155.00		

15. DUCTWORK

SPECIFICATIONS		UNIT	PRICE	LOCAL AREA MODIFICATION	
				%	PRICE
REGISTER	CUT OUT AND INSTALL REGISTER IN EXISTING WARM AIR OR RETURN DUCT	EA	180.00		
DUCTWORK	EXTEND SUPPLY OR RETURN DUCT TO ROOM ADDITION WITHIN 10 FEET	EA	543.00		
	EXTEND DUCTWORK IN BASEMENT	LF	16.20		
	CONVERT EXISTING LOW DUCT TO HIGH-LOW DUCT AND REGISTER FOR CENTRAL AIR CONDITIONING, **NOT** INCLUDING WALL REPAIR OR PLASTERING	EA	294.00		
	RUN DUCTWORK FROM WALL TO BASE OF EITHER KITCHEN CABINET OR VANITY WITH NEW REGISTER	EA	219.00		
	EXTEND HIGH REGISTER OUT TO SOFFIT IN KITCHEN	EA	165.00		
HUMIDIFIER	INSTALL FLUSH TYPE HUMIDIFIER ON EXISTING FURNACE DUCT SYSTEM	EA	549.00		

15

SPECIFICATIONS		UNIT	PRICE	LOCAL AREA MODIFICATION %	LOCAL AREA MODIFICATION PRICE
	THE PRICES SHOWN IN THIS SECTION INCLUDE AN ELECTRICAL SUBCONTRACTOR'S OVERHEAD AND PROFIT				
SERVICE	INCREASE SERVICE WITH CIRCUIT BREAKERS				
	TO 150 AMPS	EA	804.00		
	TO 200 AMPS	EA	909.00		
	INCREASE SERVICE WITH FUSE BOX				
	TO 150 AMPS	EA	570.00		
	TO 200 AMPS	EA	690.00		
	RE-LOCATE EXISTING SERVICE ON SAME SERVICE PANEL	EA	510.00		
FUSE	INSTALL FUSE BOX				
	2-CIRCUIT	EA	129.00		
	4-CIRCUIT	EA	165.00		
	6-CIRCUIT	EA	192.00		
	8-CIRCUIT	EA	216.00		
SPECIAL WIRING	SINGLE HEAVY-DUTY LINE FOR APPLIANCE OR WORKSHOP				
	BASEMENT	EA	129.00		
	FIRST FLOOR	EA	159.00		
	SECOND FLOOR	EA	204.00		
	THIRD FLOOR	EA	279.00		
	REMOVE AND RE-INSTALL CEILING LIGHT FIXTURE ON ONE TRIP	EA	42.00		
	REWIRE WITH:				
	ROMEX CABLE	EA	108.00		
	BX ARMORED CABLE	EA	135.00		
	EA = EACH OUTLET				

16. ELECTRICAL OUTLETS

SPECIFICATIONS		UNIT	PRICE	LOCAL AREA MODIFICATION %	LOCAL AREA MODIFICATION PRICE
	NOTE: ALL ELECTRICAL OUTLETS SHOWN HERE ARE FOR NEW WORK (IN OPEN WALLS AND CEILINGS) AND INCLUDE CIRCUITS AS REQUIRED				
DUPLEX OUTLETS	DUPLEX OUTLETS	EA	57.00		
	WEATHERPROOF OUTLET	EA	105.00		
APPLIANCE OUTLETS	DISPOSAL OUTLET WITH SWITCH	EA	132.00		
	DISHWASHER OUTLET	EA	102.00		
	HOOD AND FAN OUTLET	EA	105.00		
	APPLIANCE OUTLET	EA	93.00		
GFIC	GROUND FAULT OUTLET	EA	117.00		
	GROUND FAULT BREAKER	EA	90.00		
	TRASH COMPACTOR OUTLET	EA	93.00		
220-VOLT OUTLET	220-VOLT OUTLET FOR A/C, CLOTHES DRYER, RANGE OR OVEN				
	BASEMENT	EA	156.00		
	FIRST FLOOR	EA	213.00		
	SECOND FLOOR	EA	255.00		
	THIRD FLOOR	EA	303.00		
SWITCH	SINGLE-POLE SWITCH	EA	57.00		
	THREE-WAY SWITCH, SET	EA	111.00		
	REPLACE EXISTING SWITCH WITH DIMMER SWITCH	EA	57.00		
FISH OUTLETS OR SWITCHES	FISH OUTLETS OR SWITCHES IN CLOSED WALLS AND CEILINGS, **ADD**				
	FRAME WALL	EA	39.00		
	MASONRY WALL	EA	75.00		

SPECIFICATIONS		UNIT	PRICE	LOCAL AREA MODIFICATION %	LOCAL AREA MODIFICATION PRICE
CEILING FIXTURE	CEILING OUTLET AND SWITCH (IN ADDITION TO COST OF FIXTURE)	EA	111.00		
FLUOR-ESCENT BATH FIXTURE	BATH FIXTURE OVER MEDICINE CABINET, ONE TUBE AND ONE PLUG, INCLUDING OUTLET, FIXTURE @ $40 AND SWITCH	EA	198.00		
RECESSED FIXTURE	RECESSED 6" ROUND FIXTURE @ $30, INCLUDING FIXTURE, OUTLET AND SWITCH	EA	204.00		
FLUOR-ESCENT FIXTURE	FLUORESCENT LIGHT, INCLUDING FIXTURE, OUT-LET AND SWITCH, OPEN FRAMING				
	4'-0" (2 TUBE) @ $50	EA	213.00		
	4'-0" (4 TUBE) @ $75	EA	249.00		
	INSTALL FLUORESCENT 12" FIXTURE @ $20 UNDER CABINET IN KITCHEN WITH SWITCH ON FIXTURE	EA	99.00		
SPOTLIGHT	INSTALL SWITCH AND OUTSIDE SPOTLIGHTS @ $25, TWO SPOTLIGHTS ON EAVES OF HOUSE				
	FIRST FLOOR	EA	267.00		
	SECOND FLOOR	EA	306.00		
EXTERIOR FIXTURE	INSTALL SWITCH AND OUTSIDE OUTLET OVER DOOR OR AT SIDE OF DOOR (IN ADDITION TO COST OF FIXTURE)	EA	150.00		

16. KITCHEN & BATHROOM FANS AND HOODS

SPECIFICATIONS		UNIT	PRICE	LOCAL AREA MODIFICATION	
				%	PRICE
	NOTE: ALL FANS & HOODS SHOWN HERE ARE FOR NEW WORK (IN OPEN WALLS AND CEILINGS) AND INCLUDE CIRCUITS AS REQUIRED				
BATHROOM FANS	EXHAUST FAN FOR BATHROOM @ $40, INCLUDING SWITCH	EA	204.00		
	1,000 WATT COMBINATION FAN & HEATER @ $95 IN BATHROOM, INCLUDING SWITCH	EA	291.00		
	COMBINATION FAN, HEATER AND LIGHT @ $120 IN BATHROOM, INCLUDING SWITCH	EA	324.00		
	EXHAUST FAN AND LIGHT @ $55 IN BATHROOM, INCLUDING SWITCH	EA	231.00		
RANGE HOODS	DUCT TYPE				
	30" @ $70	EA	279.00		
	36" @ $75	EA	285.00		
	42" @ $77	EA	288.00		
	DUCTLESS				
	30" @ $98	EA	297.00		
	36" @ $106	EA	309.00		
	42" @ $109	EA	315.00		
	COOKTOP DOWNDRAFT DUCTED EXHAUST FAN	EA	279.00		
DUCTWORK	METAL DUCTWORK FOR HOOD OR FAN INSTALLATION, **NO** WALL BREAKTHROUGH	LF	21.00		
	BREAK THROUGH WALL AND VENT FAN				
	FRAME WALL	EA	42.00		
	MASONRY WALL	EA	78.00		
	VENT FAN THROUGH ROOF OF SINGLE-STORY HOUSE	EA	69.00		
CEILING FAN	INSTALL CEILING FAN USING EXISTING WIRING	EA	294.00		
	INSTALL LIGHT KIT FOR CEILING FAN	EA	60.00		

16

16. HOUSE EXHAUST FANS, CENTRAL VACUUM, SMOKE DETECTOR

SPECIFICATIONS		UNIT	PRICE	LOCAL AREA MODIFICATION	
				%	PRICE
ATTIC EXHAUST FAN	• 1100-1500 CFM • INCL. ELECTRICAL WIRING • GABLE VENT MOUNTED				
	ATTIC EXHAUST FAN WITH AUTOMATIC THERMO-STAT @ $64	EA	288.00		
	SAME AS ABOVE, ROOF MOUNTED FAN AND ROOF HOOD				
	Type of Roof				
	BUILT-UP	EA	495.00		
	ASPHALT SHINGLES, ROLL ROOFING OR SELVAGE	EA	345.00		
	ASBESTOS, SLATE OR METAL	EA	435.00		
	CEDAR SHINGLES OR SHAKES	EA	375.00		
	WHOLE HOUSE CEILING MOUNTED EXHAUST FAN	EA	489.00		
CENTRAL VACUUM SYSTEM	INCLUDING POWER UNIT, HOSE, CLEANING TOOLS, TUBING INSTALLATION EA = CENTRAL UNIT AND THREE STATIONS	EA	1544.00		
	EACH ADDITIONAL STATION **ADD**	EA	270.00		
SMOKE DETECTOR	HARD-WIRED AUTOMATIC FIRE AND SMOKE DE-TECTOR	EA	165.00		

16

17. INSULATION, NEW WORK

SPECIFICATIONS			UNIT	PRICE	LOCAL AREA MODIFICATION %	LOCAL AREA MODIFICATION PRICE
	NOTE: THE PRICES ON THIS PAGE ARE FOR NEW WORK ONLY					
FIBERGLASS BLANKET	STAPLED TO OPEN FRAMING OR LAID FLAT BETWEEN CEILING JOISTS ON INSTALLED DRYWALL					
	UNFACED	3-1/2"	SF	.66		
		6"	SF	.84		
		9"	SF	1.16		
	FOILBACK ONE FACE					
		3-1/2	SF	.72		
		6"	SF	.92		
		9"	SF	1.19		
	KRAFTBACK ONE FACE					
		3-1/2"	SF	.68		
		6"	SF	.86		
		9"	SF	1.13		
	STAPLED TO FURRING STRIPS					
	UNFACED	1"	SF	.54		
	FOILBACK ONE FACE	1"	SF	.59		
	STAPLED TO BASEMENT CEILING					
	UNFACED	6"	SF	.84		
	FOILBACK ONE FACE	3-1/2"	SF	.71		
	RIGID INSULATION STAPLED TO ROOF RAFTERS OF CATHEDRAL CEILING					
	FOIL FACED	3-1/2"	SF	.72		
		6"	SF	.92		
	1/2" FOAM INSULATION BOARD		SF	.93		
PERIMETER INSULATION	FIBERBOARD, STYROFOAM OR RUBBERBOARD, SIDE OF EXTERIOR SLAB AND 18" IN FROM EDGE AT BOTTOM					
		3/4"	LF	1.41		
		1"	LF	1.53		
		2"	LF	2.34		
	LF = PERIMETER OF SLAB					

17

SPECIFICATIONS		UNIT	PRICE	LOCAL AREA MODIFICATION	
				%	PRICE
	NOTE: THE PRICES ON THIS PAGE ARE FOR INSTALLATIONS IN OLD WORK				
FIBERGLASS BLANKET	STAPLED TO EXISTING KNEEWALLS IN ATTIC				
	UNFACED (R-11) 3-1/2	SF	.75		
	FOIL FACE (INSIDE) 3-1/2	SF	.80		
	KRAFT FACED 3-1/2"	SF	.77		
	FIBERGLASS BLANKET TO FILL SOFFITS AND DROPPED CEILINGS IN ATTICS CF = CUBIC FEET	CF	2.42		
	STAPLE 3-1/2" FIBERGLASS BLANKET TO TRAP DOOR IN ATTIC OR TO BACK OF PULLDOWN STAIRS	EA	12.75		
FIBERGLASS BATTS	UNFACED, INSTALLED OVER EXISTING INSULATION BEHIND KNEEWALLS, (R-8) 2-1/2"	SF	.62		
BAFFLE	INSTALL BAFFLE AROUND RECESSED LIGHTS				
	OPEN CEILING	EA	18.75		
	RESTRICTED CEILING	EA	23.25		
	INSTALL BAFFLE AROUND CHIMNEY	EA	33.00		
	INSTALL BAFFLE AROUND FAN	EA	18.75		
	INSTALL VENTILATION BAFFLES IN RAFTER SPACES	LF	.80		

17. INSULATION, BLOWN IN

SPECIFICATIONS		UNIT	PRICE	LOCAL AREA MODIFICATION	
				%	PRICE
	NOTE: THE PRICES SHOWN ON THIS PAGE INCLUDE AN INSULATION SUBCONTRACTOR'S OVERHEAD AND PROFIT				
BLOWN-IN FIBERGLASS	BEHIND ASBESTOS, ALUMINUM OR WOOD SIDING, INCLUDING DRILLING AND PLUGGING, **NO** PAINTING, FROM OUTSIDE	SF	3.09		
	BEHIND STUCCO OR FRAME WALL, FROM OUTSIDE	SF	3.30		
	BEHIND DRYWALL OR PLASTER WALL, FROM INSIDE	SF	3.68		
	BLOWN IN BETWEEN OPEN JOISTS, UNRESTRICTED				
	R-11	SF	.72		
	R-19	SF	1.20		
	R-30	SF	1.80		
	R-38	SF	2.40		
BLOWN-IN CELLULOSE	FIRE-RETARDANT CELLULOSE BLOWN IN BETWEEN OPEN JOISTS, UNRESTRICTED				
	R-11	SF	.53		
	R-19	SF	.83		
	R-30	SF	1.35		
	R-38	SF	1.88		

17

SPECIFICATIONS		UNIT	PRICE	LOCAL AREA MODIFICATION %	LOCAL AREA MODIFICATION PRICE
RESTRICTED AREAS	IN RESTRICTED AREAS, FOR DIFFICULT ACCESS **ADD**	SF	.38		
MISCELLANEOUS	LAY 4-MIL VAPOR BARRIER ON GROUND IN CRAWL SPACE	SF	.20		
	WRAP HEATING/COOLING PIPES OR DUCTS IN CRAWL SPACE OR ATTIC WITH INSULATION				
	1/2" ON PIPES	LF	.90		
	1-1/2" ON DUCTS	LF	1.05		
RADIANT BARRIERS	REINFORCED FOIL HEAT SHIELD (FOIL FACES DOWN)	SF	.32		

18. GYPSUM DRYWALL, NEW WORK

SPECIFICATIONS		UNIT	PRICE	LOCAL AREA MODIFICATION	
				%	PRICE
GYPSUM DRYWALL ON NEW WALL -- COMPLETE JOB	*UP TO 300 SF* OF WALL, NAILED, TAPED, FINISHED AND SANDED, 3 COATS				
	3/8"	EA PLUS SF	150.00 .77		
	1/2"	EA PLUS SF	150.00 .81		
	1/2" FIRECODE	EA PLUS SF	150.00 .89		
	5/8" FIRECODE	EA PLUS SF	150.00 .95		
	1/2" MOISTURE RESISTANT	EA PLUS	150.00		
	EA = EACH JOB (**NOT** EACH SHEET)	SF	.87		
	SAME AS ABOVE, WHEN TOTAL JOB IS *OVER 300 SF*				
	3/8"	SF	1.26		
	1/2"	SF	1.31		
	1/2" FIRECODE	SF	1.38		
	5/8" FIRECODE	SF	1.46		
	1/2" MOISTURE RESISTANT	SF	1.37		
NAILED ONLY	NAILED ONLY, APPLIED TO STUDS OR FURRING				
	3/8"	SF	.51		
	1/2"	SF	.56		
	1/2" FIRECODE	SF	.63		
	5/8" FIRECODE	SF	.69		
	1/2" MOISTURE RESISTANT	SF	.62		
TAPE AND FINISH ONLY	TAPE JOINTS, FINISH AND SAND ONLY				
	UP TO 300 SF OF WALL	EA	93.00		
	OVER 300 SF OF WALL	SF	.77		
	1/2" BLUEBOARD	SF	.78		
	SKIM COAT PLASTER	SF	.80		

18

SPECIFICATIONS		UNIT	PRICE	LOCAL AREA MODIFICATION %	LOCAL AREA MODIFICATION PRICE
GYPSUM DRYWALL OVER EXISTING WALL – COMPLETE JOB	*UP TO 300 SF* OF WALL, NAILED WITH NAILS OF REQUIRED LENGTH THROUGH EXISTING WALL FINISH AND INTO STUDS, TAPED, FINISHED AND SANDED				
	3/8"	EA PLUS SF	111.00 1.04		
	1/2"	EA PLUS SF	111.00 1.08		
	5/8" FIRECODE	EA PLUS	111.00		
	EA = EACH JOB (**NOT** EACH SHEET)	SF	1.25		
	SAME AS ABOVE, WHEN TOTAL JOB IS *OVER 300 SF*				
	3/8"	SF	1.01		
	1/2"	SF	1.10		
	5/8" FIRECODE	SF	1.29		
TEXTURE SPACKLE FINISH	SPRAY FINISHED DRYWALL WITH SPACKLE OR TEXTURE SPRAY, *UP TO 300 SF* OF WALL	EA PLUS SF	63.00 .48		
	SAME AS ABOVE WHEN JOB IS *OVER 300 SF*	SF	.68		
NAILED OVER EXISTING WALL	DRYWALL OVER EXISTING PLASTERED WALL WITH NAILS OF REQUIRED LENGTH THROUGH EXISTING PLASTER & LATH & INTO STUDS, NAILED ONLY				
	3/8"	SF	1.01		
	1/2"	SF	1.05		
	5/8" FIRECODE	SF	1.22		
VINYL SURFACE DRYWALL	VINYL SURFACE GYPSUM PANELS WITH COLOR PINS	SF	1.91		

18

18. LATHING

SPECIFICATIONS		UNIT	PRICE	LOCAL AREA MODIFICATION	
				%	PRICE
GYPSUM LATH	GYPSUM LATH NAILED TO WOOD STUDS, INCLUDING CORNER BEADS AND CORNERITE				
	3/8"	SF	.84		
	1/2"	SF	.89		
METAL LATH	3.4 DIAMOND MESH NAILED TO WOOD STUDS	SF	1.04		
WOOD LATH	3/8" X 1-1/2" X 48" LATH NAILED TO STUDS (3/8" SPACED)	SF	1.53		
GROUNDS	WOOD GROUNDS FOR PLASTERING				
	1" X 1"	LF	.89		
	1" X 3"	LF	.98		

18

SPECIFICATIONS		UNIT	PRICE	LOCAL AREA MODIFICATION %	LOCAL AREA MODIFICATION PRICE
	THE PRICES SHOWN ON THIS PAGE INCLUDE A PLASTERING SUBCONTRACTOR'S OVERHEAD AND PROFIT				
PLASTER	GYPSUM PLASTER OVER EXISTING GYPSUM LATH				
	2 COATS	SF	1.95		
	3 COATS	SF	2.57		
	3 COATS GYPSUM PLASTER				
	OVER MASONRY	SF	3.11		
	OVER METAL LATH	SF	2.81		
	OVER WOOD LATH	SF	2.66		
SKIM-COAT PLASTERING	TAPE GYPSUM BOARD (SHEETROCK) AND APPLY LIME PUTTY PLASTER FINISH	SF	1.25		

18. CERAMIC TILE

SPECIFICATIONS		UNIT	PRICE	LOCAL AREA MODIFICATION	
				%	PRICE
IN MUD	INCLUDING METAL LATHING AND SCRATCH COAT, WITH AVERAGE AMOUNT OF COVE AND BULL-NOSE, 4-1/4" X 4-1/4"	SF	18.23		
WONDER-BOARD OR DUROCK	CERAMIC TILE BACKER BOARD, 1/2" 3' X 5' SHEETS NOTE: CERAMIC TILE MAY ALSO BE INSTALLED OVER MOISTURE-RESISTANT DRYWALL; SEE PAGE 224	SF	6.60		
THIN-SET OR MASTIC	CERAMIC TILE WITH AVERAGE AMOUNT OF COVE AND BULLNOSE, 4-1/4" X 4-1/4", 6" X 6", AND 8" X 8", OVER EXISTING BACKER BOARD OR MOISTURE RESISTANT DRYWALL	SF	11.55		
FIXTURES	CERAMIC TILE BATHROOM FIXTURES: TOWEL BAR, TOOTHBRUSH HOLDER, PAPER HOLDER, SOAP AND GRAB	SET	63.00		
EXTEND TILE	EXTEND CERAMIC TILE ABOVE EXISTING TILE WALL				
	IN MUD	SF	18.83		
	IN MASTIC	SF	17.90		

18

SPECIFICATIONS		UNIT	PRICE	LOCAL AREA MODIFICATION %	LOCAL AREA MODIFICATION PRICE
PLYWOOD PANELING	PREFINISHED, 1/4" X 4 X 8 SHEETS NAILED TO STUDS OR FURRING. **NO** MOULDING INCLUDED				
	Retail Price Per 4 X 8 Sheet				
	10.00	SF	1.35		
	11.00	SF	1.40		
	12.00	SF	1.46		
	13.00	SF	1.50		
	14.00	SF	1.55		
	15.00	SF	1.59		
	16.00	SF	1.64		
	17.00	SF	1.68		
	18.00	SF	1.73		
	19.00	SF	1.77		
	20.00	SF	1.82		
	25.00	SF	2.04		
	30.00	SF	2.31		
	35.00	SF	2.55		
	40.00	SF	2.81		
	45.00	SF	3.08		
V-JOINT, SOLID WOOD	UNFINISHED, V-JOINT 3/4" THICK AND RANDOM WIDTHS TO 10"				
	INCLUDING INSTALLATION OF 1 X 4 BASE OF SAME MATERIAL TO WHICH THE PANELING IS BUTTED				
	KNOTTY PINE	SF	4.58		
	CEDAR	SF	4.62		
	REDWOOD	SF	5.70		
CEDAR CLOSET LINING	3/8" T&G AND END MATCHED, APPLIED OVER EXISTING SOLID WALL	SF	4.71		
	1/4" CEDAR PARTICLEBOARD, 4-0 X 8-0 SHEETS	SF	2.64		

18. PATCHING WALLS

SPECIFICATIONS		UNIT	PRICE	LOCAL AREA MODIFICATION %	LOCAL AREA MODIFICATION PRICE
	THE PRICES SHOWN ON THIS PAGE INCLUDE CERAMIC TILE, DRYWALL AND PLASTERING SUBCONTRACTORS' OVERHEAD AND PROFIT				
DRYWALL PATCHING	PATCH HOLES AND/OR CRACKS IN DRYWALL EA = EACH JOB	EA PLUS SF	180.00 2.25		
CORRECT NAIL POPS	CORRECT NAIL POPS IN EXISTING WALL, **NO** PAINT TOUCH-UP INCLUDED EA = UP TO ONE DOZEN NAIL POPS	EA	27.00		
PLASTER PATCHING	PATCH HOLES AND/OR CRACKS IN EXISTING PLASTER, INCLUDING LATHING WHERE REQUIRED EA = EACH JOB	EA PLUS SF	285.00 2.40		
PATCHING	PATCH CERAMIC TILE WALL, INCLUDING REMOVING DEFECTIVE OR BROKEN TILES EA = EACH JOB	EA PLUS SF	99.00 19.95		
	PATCH CERAMIC TILE COVE, BASE OR BULLNOSE	LF	13.16		
	COVER EXISTING CERAMIC TILE WALL WITH URETHANE FINISH	SF	8.25		

18

SPECIFICATIONS		UNIT	PRICE	LOCAL AREA MODIFICATION %	LOCAL AREA MODIFICATION PRICE
GYPSUM DRYWALL ON NEW CEILING	*UP TO 300 SF* CEILING, APPLIED TO JOISTS OR FURRING, NAILED, TAPED, FINISHED AND SANDED, 3 COATS				
	3/8"	EA	150.00		
		PLUS SF	.83		
	1/2"	EA	150.00		
		PLUS SF	.87		
	1/2" FIRECODE	EA	150.00		
		PLUS SF	.90		
	5/8" FIRECODE	EA	150.00		
		PLUS SF	1.01		
	1/2" MOISTURE RESISTANT	EA	150.00		
	EA = EACH JOB (**NOT** EACH SHEET)	PLUS SF	.93		
	SAME AS ABOVE, WHEN TOTAL JOB IS *OVER 300 SF*				
	3/8"	SF	1.32		
	1/2"	SF	1.37		
	1/2" FIRECODE	SF	1.40		
	5/8" FIRECODE	SF	1.52		
	1/2" MOISTURE RESISTANT	SF	1.43		
	FOR CEILINGS OVER 8'-0" HIGH **ADD**	SF	.17		
NAILED ONLY	NAILED ONLY, APPLIED TO JOISTS OR FURRING				
	3/8"	SF	.54		
	1/2"	SF	.59		
	5/8" FIRECODE	SF	.72		
	FOR CEILINGS OVER 8'-0" HIGH **ADD**	SF	.08		
TAPE AND FINISH ONLY	TAPE JOINTS, FINISH AND SAND ONLY				
	UP TO 300 SF CEILING	EA	93.00		
	OVER 300 SF CEILING	SF	.80		
	FOR CEILINGS OVER 8'-0" HIGH **ADD**	SF	.11		
	1/2" BLUEBOARD	SF	.78		
	SKIM COAT PLASTER	SF	.80		

19

19. DRYWALL CEILING, OVER EXISTING

SPECIFICATIONS		UNIT	PRICE	LOCAL AREA MODIFICATION %	LOCAL AREA MODIFICATION PRICE
GYPSUM DRYWALL OVER EXISTING CEILING	*UP TO 300 SF* CEILING, INSTALL DRYWALL OVER EXISTING PLASTERED CEILING WITH NAILS OF REQUIRED LENGTH THROUGH EXISTING PLASTER & LATH & INTO JOISTS, TAPED, FINISHED AND SANDED, 3 COATS				
	3/8"	EA PLUS SF	150.00 1.13		
	1/2"	EA PLUS SF	150.00 1.20		
	5/8" FIRECODE	EA PLUS	150.00		
	EA = EACH JOB (**NOT** EACH SHEET)	SF	1.38		
	SAME AS ABOVE, WHEN TOTAL JOB IS *OVER 300 SF*				
	3/8"	SF	1.62		
	1/2"	SF	1.70		
	5/8" FIRECODE	SF	1.89		
	FOR CEILINGS OVER 8'-0" HIGH **ADD**	SF	.17		
TEXTURE SPACKLE FINISH	SPRAY FINISHED DRYWALL WITH SPACKLE OR TEXTURE SPRAY, *UP TO 300 SF* OF CEILING	EA PLUS SF	63.00 .50		
	SAME AS ABOVE WHEN JOB IS *OVER 300 SF*	SF	.69		
CORRECT NAIL POPS	CORRECT NAIL POPS IN EXISTING CEILING. **NO** PAINT TOUCH-UP INCLUDED EA = DOZEN NAIL POPS	EA	27.00		
DRYWALL PATCHING	PATCH HOLES AND/OR CRACKS IN DRYWALL EA = EACH JOB	EA PLUS SF	180.00 2.25		

19

SPECIFICATIONS		UNIT	PRICE	LOCAL AREA MODIFICATION	
				%	PRICE
GYPSUM LATH	GYPSUM LATH NAILED TO WOOD STUDS, INCLUDING CORNERITE				
	3/8"	SF	.84		
	1/2"	SF	.89		
METAL LATH	3.4 DIAMOND MESH NAILED TO JOISTS OR FURRING	SF	1.02		
WOOD LATH	3/8" X 1-1/2" X 48" WOOD LATH NAILED TO JOISTS (3/8" SPACED)	SF	1.64		

19. PLASTER CEILING

SPECIFICATIONS		UNIT	PRICE	LOCAL AREA MODIFICATION %	LOCAL AREA MODIFICATION PRICE
PLASTER	GYPSUM PLASTER OVER EXISTING GYPSUM LATH				
	2 COATS	SF	1.95		
	3 COATS	SF	2.57		
	PLASTER, 3 COATS OVER METAL LATH	SF	2.81		
	OVER WOOD LATH	SF	2.66		
PATCHING	PATCH HOLES AND/OR CRACKS IN EXISTING PLASTER, INCLUDING LATHING WHERE REQUIRED EA = EACH JOB	EA PLUS SF	285.00 2.40		
	PATCH CEILING WITH DRYWALL OR PLASTER WHERE WALL BELOW HAS BEEN REMOVED EA = EACH JOB LF = LENGTH OF WALL REMOVED	EA PLUS LF	30.00 4.50		

19

SPECIFICATIONS		UNIT	PRICE	LOCAL AREA MODIFICATION %	LOCAL AREA MODIFICATION PRICE
	12" X 12" OR 12" X 24" TILE OVER EXISTING CEILING OR FURRING STRIPS WITH STAPLES OR ADHESIVE ***Tile SF Retail Cost***				
	.35	SF	1.41		
	.40	SF	1.50		
	.45	SF	1.58		
	.50	SF	1.65		
	.55	SF	1.73		
	.60	SF	1.82		
	.70	SF	1.97		
	.80	SF	2.13		
FURRING	INSTALL 1 X 3 FURRING STRIPS 12" O.C., READY FOR INSTALLATION OF CEILING TILE	SF	1.16		
	IF FURRING NAILED THROUGH EXISTING PLASTERED CEILING	SF	1.05		
	SUSPEND 1 X 3 FURRING STRIPS MORE THAN 4" BELOW JOISTS	SF	2.69		
GRID SYSTEM	SUSPEND CEILING WITH STEEL GRID SYSTEM USING 24" X 48" SECTIONS, INCLUDING ENTIRE GRID SYSTEM AND CEILING PANELS ***Ceiling Panels SF Retail Cost***				
	.50	SF	3.50		
	.55	SF	3.57		
	.60	SF	3.65		
	.65	SF	3.72		
	.70	SF	3.80		
	.75	SF	3.87		
	.80	SF	3.95		

20. INTERIOR MOULDINGS

SPECIFICATIONS		UNIT	PRICE	LOCAL AREA MODIFICATION	
				%	PRICE
BASE	PINE CLAM, RANCH OR COLONIAL BASE				
	2"	LF	1.68		
	2-1/4"	LF	1.98		
	3-1/4"	LF	2.48		
	4-1/4"	LF	3.24		
	5-1/4"	LF	3.89		
	8"	LF	6.74		
	OAK BASE, 3-1/4"	LF	4.80		
SHOE MOULD	1/2" X 3/4"				
	HARDWOOD	LF	1.88		
	PINE	LF	1.10		
3-PIECE BASE	• TOP MOULD, 11/16" X 1-1/8" • OAK SHOE 1/2" X 3/4" • BASE 11/16" X 3-1/2" LF = ALL THREE PIECES	LF	5.43		
CEILING MOULDING	BED, COVE, CROWN OR PICTURE MOULDS				
	3/4"	LF	1.76		
	1-5/8"	LF	2.28		
	2-1/4"	LF	2.70		
	3-1/4"	LF	3.78		
	4-1/4"	LF	4.32		
	5-1/4"	LF	5.09		

20

SPECIFICATIONS		UNIT	PRICE	LOCAL AREA MODIFICATION %	LOCAL AREA MODIFICATION PRICE
3-MEMBER WOOD CORNICE	INSTALLED ON PERIMETER OF ROOM OVER TRUE WALLS 5/8 X 1-3/4" AND 3/4" X 4-3/4" AND 5/8" X 1-3/4" LF = INCLUDES ALL THREE PIECES	LF	9.05		
CHAIR RAIL	ONE PIECE				
	5/8" X 2-1/2"	LF	2.88		
	5/8" X 3-1/2"	LF	3.72		
	TWO-PIECE ROUND-EDGE CHAIR RAIL WITH CAP TRIM	LF	5.39		
PANEL STRIPS	PANEL STRIPS NAILED TO EXISTING PLASTERED DRYWALL OR PLYWOOD WALL FOR PANEL EFFECT, ALL JOINTS MITERED				
	5/8" X 7/8"	LF	1.79		
	1-1/16" X 2-1/4"	LF	2.43		

20. SPECIAL MOULDINGS AND PLASTER MOULD

SPECIFICATIONS		UNIT	PRICE	LOCAL AREA MODIFICATION	
				%	PRICE
SPECIAL WOOD MOULDINGS	CUSTOM MOULDINGS MILLED TO ORDER IN SMALL QUANTITIES. MATERIALS COSTS INCLUDE SETTING UP AND SHAPING IN SHOP, & LABOR COST IS FOR INSTALLATION ON JOB	EA PLUS	114.00		
	3/4" X 1-1/8"	LF	2.07		
	2-1/4"	LF	2.67		
	3-1/2"	LF	3.17		
	4-1/2"	LF	3.68		
	5-1/2"	LF	3.99		
	FOR SPECIAL MOULDINGS REQUIRING NEW BIT **ADD**	EA	84.00		
PLASTER MOULD	ORNAMENTAL PLASTER MOULDING, STOCK OR SCULPTED, INSTALLED ON JOB BY ARCHITECTURAL SCULPTORS				
	Projection ***Depth***				
	2" 4"	LF	12.30		
	2" 7"	LF	13.13		
	3" 4"	LF	13.43		
	3" 8"	LF	14.10		
	4" 5"	LF	17.18		
	5" 7"	LF	28.35		
	SAME AS ABOVE, ON CURVED WALLS **ADD**	LF	25%		

20

SPECIFICATIONS		UNIT	PRICE	LOCAL AREA MODIFICATION	
				%	PRICE
FIREPLACE MANTEL	PINE WITH PLAIN OR FLUTED SIDES, PRE-ASSEMBLED				
	OVERALL WIDTH UP TO 72", SHELF SIZE UP TO 12", WOOD OPENING UP TO 53" IN WIDTH				
	ECONOMY	EA	342.00		
	MEDIUM	EA	672.00		
	PREMIUM	EA	1176.00		
	PLAIN PINE SURROUND MOULDING WITH NO SHELF, SURROUND MOULDING UP TO 1-5/8" X 5-3/4"	EA	159.00		
CHINA CORNER CASE	PINE, PRE-ASSEMBLED, OVERALL HEIGHT TO 7-4, OVERALL WIDTH TO 2-6				
	ECONOMY	EA	456.00		
	MEDIUM	EA	540.00		
	PREMIUM	EA	666.00		
CASED OPENING	JAMBS AND TWO SIDES OF 2-1/2" DOOR TRIM				
	3-0 X 6-8	EA	114.00		
	6-0 X 6-8	EA	129.00		
TRIM EXISTING DOOR OPENING	ONE SIDE ONLY, 2-1/2" DOOR TRIM ON EXISTING JAMBS				
	3-0 X 6-8	EA	42.00		
	6-0 X 6-8	EA	45.00		
TRIM EXISTING WINDOW	INTERIOR TRIM ONLY • 2-1/2" WINDOW TRIM • APRON • STOOL CAP • WINDOW STOPS				
	Window Size				
	2-0 X 3-2	EA	63.00		
	3-0 X 5-2	EA	75.00		

20. BOOKSHELVES

SPECIFICATIONS		UNIT	PRICE	LOCAL AREA MODIFICATION %	LOCAL AREA MODIFICATION PRICE
BUILT-IN BOOKCASE	• NAILED AND GLUED ENDS • VERTICAL SUPPORTS 30" O.C. AND AT EACH END • KICKBOARD • TOP RAIL • SHELVES SPACED 10" O.C. VERTICALLY MEASURED • BUILD AND INSTALL SF = FRONT SQUARE FOOTAGE				
	TIGHT KNOT PINE SHELVING				
	8"	SF	7.52		
	10"	SF	8.57		
	12"	SF	9.65		
	CLEAR WHITE PINE				
	8"	SF	15.24		
	10"	SF	18.24		
	12"	SF	21.30		
	OAK				
	8"	SF	17.18		
	10"	SF	20.72		
	12"	SF	24.24		
	BIRCH PLYWOOD (3/4")				
	8"	SF	9.96		
	10"	SF	11.54		
	12"	SF	13.13		
	16"	SF	16.34		
SHELF MOULDING	MOULDING AT SHELF EDGE	LF	1.91		
STILES	1 X 2 CWP STILES ALONG VERTICAL SUPPORTS OF BOOKCASE	LF	1.44		
BACKING	1/4" PLYWOOD OR HARDBOARD	SF	2.10		

20

SPECIFICATIONS		UNIT	PRICE	LOCAL AREA MODIFICATION %	LOCAL AREA MODIFICATION PRICE
PIN TYPE SHELF SUPPORTS	• DRILL 1/4" HOLES IN SIDES OF VERTICAL SUPPORTS • INSERT METAL PIN TYPE SHELF SUPPORTS • INSTALL SHELVES ON SHELF SUPPORTS WITHOUT NAILING **ADD**	SF	.63		
ADJUSTABLE STANDARDS AND 3/4" SUPPORTS	• FASTEN TWO 1/2" ADJUSTABLE STANDARDS ALONG SIDES OF VERTICAL SUPPORTS • INSERT 3/4" SUPPORTS IN SLOTS IN ADJUSTABLE STANDARDS • INSTALL SHELVES ON SHELF SUPPORTS WITHOUT NAILING **ADD**	SF	1.17		
OPEN SHELVES ON BRACKETS	• FASTEN SLOTTED STANDARDS TO EXISTING WALL • INSERT SHELF BRACKETS UP TO 12" LONG INTO SLOTTED STANDARDS • INSTALL SHELVES ON BRACKETS WITHOUT NAILING • **NO** VERTICAL WOOD SUPPORTS				
	#2 PINE 10"	LF	8.72		
	12"	LF	9.74		
	CLEAR WHITE PINE 10"	LF	12.77		
	12"	LF	14.43		
	OAK 10"	LF	12.27		
	12"	LF	13.91		
	BIRCH PLYWOOD 10"	LF	10.49		
	12"	LF	11.45		
	LF = TOTAL LENGTH OF ALL SHELVES				

20

20. CLOSET TRIM

SPECIFICATIONS		UNIT	PRICE	LOCAL AREA MODIFICATION	
				%	PRICE
CLOTHES CLOSET	24" TO 30" DEEP, FULLY TRIMMED AS FOLLOWS: • #2 PINE BASE AND SHOE • #2 PINE HOOKSTRIP • 12" PARTICLEBOARD SHELF • 1-3/8" CLOTHES POLE AND CLOTHES POLE SOCKETS • **NO** FRAMING, WALL COVERING, DOOR OR DOOR TRIM LF = WIDTH OF CLOSET	LF	23.67		
WALK-IN CLOSET	SAME AS ABOVE, ADDING ONE POLE AND ONE SHELF	SF	6.26		
CLOSET SHELVES	ADDITIONAL STORAGE SHELVES IN ABOVE CLOSETS, #2 PINE OR PLYWOOD SHELVES SET ON WOOD CLEATS				
	12"	LF	4.85		
	16"	LF	5.81		
LINEN CLOSET	24" DEEP WITH 24" PLYWOOD OR PARTICLEBOARD SHELVES 12" O.C., VERTICALLY MEASURED, #2 PINE BASE AND #2 PINE SHELF CLEATS	LF	91.79		
BASEMENT CLOSET	BUILD CLOSET UNDER EXISTING STAIRWAY, 36" X 36" • FRAME DOORWAY • TRIM DOOR INSIDE & OUT • CLOTHES CLOSET TRIM • DRYWALL, FINISHED	EA	603.00		

20

SPECIFICATIONS		UNIT	PRICE	LOCAL AREA MODIFICATION	
				%	PRICE
ATTIC STAIRWAY	BETWEEN TWO WALLS, HOUSED OUT PINE STRINGERS, 1-1/8" PINE TREADS, 3/4" PINE RISERS, FIR HANDRAIL AND TWO METAL HANDRAIL BRACKETS, SHOP BUILT AND DELIVERED TO JOB BY STAIRBUILDER	EA	705.00		
	SAME AS ABOVE WITH RIGHT ANGLE TURN AT TOP OR BOTTOM	EA	810.00		
CELLAR STAIRWAY	STRINGERS AND TREADS OF 2 X 10 FIR, OPEN RISERS, 2 X 4 HANDRAIL	EA	567.00		
MAIN STAIRWAY	STARTING NEWEL, BIRCH HANDRAIL WITH EASEMENT AND NEWEL CAP, PINE BALUSTERS, OAK TREADS AND RISERS, SHOP BUILT AND INSTALLED ON JOB BY STAIRBUILDER				
	ONE SIDE OPEN	EA	2559.00		
	TWO SIDES OPEN	EA	3000.00		
	SAME SPECIFICATIONS AS ABOVE WITH PLATFORM AND RIGHT ANGLE TURN AT TOP OR BOTTOM **ADD**	EA	399.00		
WROUGHT IRON HANDRAIL	SUBSTITUTE WROUGHT IRON HANDRAIL IN ABOVE STAIRWAYS **ADD**				
	ONE SIDE	EA	216.00		
	TWO SIDES	EA	345.00		
WOOD HANDRAIL	2 X 2 PINE OR FIR HANDRAIL WITH STANDARD HARDWARE	LF	7.65		

20. SPIRAL/FOLDING STAIRS

SPECIFICATIONS		UNIT	PRICE	LOCAL AREA MODIFICATION %	LOCAL AREA MODIFICATION PRICE
SPIRAL STAIRWAY, WOOD	HARDWOOD TREADS AND PLATFORM, LAMINATED HANDRAIL AND CENTER POLE, BALUSTERS, UP TO 9-0 FLOOR TO FLOOR ***Diameter***				
	4-0	EA	4344.00		
	4-6	EA	4728.00		
	SAME AS ABOVE, 9-0 TO 11-0 FLOOR TO FLOOR ***Diameter***				
	4-0	EA	5334.00		
	4-6	EA	5667.00		
SPIRAL STAIRWAY, METAL	UNPAINTED METAL STAIRWAY ***Diameter***				
	5-0	EA	1647.00		
	6-0	EA	1833.00		
FOLDING STAIRWAY, WOOD	"DISAPPEARING STAIRWAY", WOOD, UP TO 10-0 FLOOR TO FLOOR X 26" WIDE, IN EXISTING FRAMED OPENING				
	ECONOMY	EA	321.00		
	MEDIUM	EA	453.00		
	PREMIUM	EA	618.00		
	OPEN CEILING AND HEAD OFF TO INSTALL FOLDING STAIRS, CUTTING ONE CEILING JOIST	EA	123.00		

20

SPECIFICATIONS		UNIT	PRICE	LOCAL AREA MODIFICATION %	LOCAL AREA MODIFICATION PRICE
STAIRS REPAIR	REPLACE BALUSTERS IN EXISTING STAIRWAY FROM FLOOR TO FLOOR, OPEN ONE SIDE, TWO BALUSTERS PER RISER				
	STOCK	PER RIS'R	34.50		
	SPECIAL MILLED	PER RIS'R	44.25		
	INSTALL NEW HANDRAIL, ONE NEWEL POST AND BALUSTERS ON EXISTING STAIRWAY, TWO BALUSTERS PER RISER	EA PLUS PER RIS'R	147.00 43.50		
	REPLACE PINE TREADS ON BASEMENT STAIRWAY, TREADS ONLY, **NO** RISERS EA = PER TREAD	EA	18.00		
RADIATOR ENCLOSURE	JOB-BUILT WITH 1 X 3 PINE STILES, BOTTOM RAIL RAISED FROM FLOOR ON FRONT AND SIDES, UP TO 12" DEEP SF = FRONT	SF	20.97		
BUILT-UP FALSE BEAM	INSTALL FALSE WOOD BOX BEAM AGAINST STRAIGHT CEILING, 2 X 4 BASE NAILER WITH 1 X 4 FACE AND 1 X 6 SIDES, 3/4" QUARTER ROUND AT CEILING INTERSECTION				
	CWP	LF	11.09		
	#2 PINE	LF	8.54		
	CLEAR RW	LF	9.06		

21. KITCHEN CABINETS, ECONOMY

SPECIFICATIONS			UNIT	PRICE	LOCAL AREA MODIFICATION %	LOCAL AREA MODIFICATION PRICE
BASE CABINETS	INSTALL PREFINISHED OR UNFINISHED KITCHEN CABINETS ON PLUMB & STRAIGHT WALLS. STAINING AND PAINTING **NOT** INCLUDED BASIC BASE CABINET INSTALLATION INCLUDES THE FOLLOWING • ONE 36" WIDE, 2 DOOR AND 2 DRAWER FRONTS, SINK BASE CABINET @ $95.00 EACH RETAIL • ONE 24" WIDE, 4 DRAWER DRAWER BASE CABINET • 1 AND 2 DOOR, 1 AND 2 DRAWER BASE CABINETS AS REQUIRED LF = TOTAL OVERALL FRONT WIDTH OF ALL BASE CABINETS		LF	114.00		
	ADDITIONAL BASE CABINETS AS REQUIRED EA = EACH CABINET					
DRAWER BASE	WITH 4 DRAWERS	15"	EA	255.00		
		18"	EA	258.00		
		21"	EA	264.00		
		24"	EA	273.00		
SINK OR BURNER BASE		30"	EA	192.00		
		36"	EA	201.00		
		42"	EA	216.00		
SINK OR BURNER FRONT	FRONT AND FLOOR (**NO** SIDES)	30"	EA	171.00		
		36"	EA	186.00		
		42"	EA	192.00		
		48"	EA	198.00		
ISLAND BASE	WITH DOORS 2 SIDES AND DRAWERS 1 SIDE	24"	EA	288.00		
		36"	EA	363.00		
		42"	EA	399.00		
		48"	EA	426.00		
CORNER BASE	(LAZY SUSAN) WITH 36" ALONG EACH WALL		EA	273.00		
BROOM CABINET	***Width*** / ***Height*** / ***Depth***					
	18" / 7-0 / 12"		EA	303.00		
	18" / 7-0 / 24"		EA	330.00		
OVEN CABINET	***Width*** / ***Height*** / ***Depth***					
	24" / 7-0 / 24"		EA	468.00		
	27" / 7-0 / 24"		EA	498.00		

SPECIFICATIONS			UNIT	PRICE	LOCAL AREA MODIFICATION	
					%	PRICE
WALL CABINETS	THE BASIC WALL CABINET INSTALLATION INCLUDES THE FOLLOWING: • TWO "ABOVE APPLIANCE" WALL CABINETS 15" HIGH AND 36" WIDE @ $76.00 EACH RETAIL • 1 AND 2 DOOR WALL CABINETS 30" HIGH AS REQUIRED LF = TOTAL OVERALL FRONT WIDTH OF ALL WALL CABINETS		LF	78.00		
	ADDITIONAL WALL CABINETS AS REQUIRED EA = EACH CABINET					
UPPER ISLAND	WITH DOORS TWO SIDES, 24" HIGH	36"	EA	279.00		
	SAME AS ABOVE, 30" HIGH	24"	EA	249.00		
		27"	EA	261.00		
		30"	EA	270.00		
		33"	EA	285.00		
		36"	EA	294.00		
		42"	EA	315.00		
		48"	EA	342.00		
CORNER WALL	(LAZY SUSAN) WITH 24" ALONG EACH WALL		EA	237.00		
MICROWAVE WALL	MICROWAVE WALL CABINET, 24"		EA	372.00		

21. KITCHEN CABINETS, BUILDER

SPECIFICATIONS			UNIT	PRICE	LOCAL AREA MODIFICATION %	LOCAL AREA MODIFICATION PRICE
BASE CABINETS	INSTALL PREFINISHED OR UNFINISHED KITCHEN CABINETS ON PLUMB & STRAIGHT WALLS. STAINING AND PAINTING **NOT** INCLUDED BASIC BASE CABINET INSTALLATION INCLUDES THE FOLLOWING • ONE 36" WIDE, 2 DOOR AND 2 DRAWER FRONTS, SINK BASE CABINET @ $160.00 EACH RETAIL • ONE 24" WIDE, 4 DRAWER DRAWER BASE CABINET • 1 AND 2 DOOR, 1 AND 2 DRAWER BASE CABINETS AS REQUIRED LF = TOTAL OVERALL FRONT WIDTH OF ALL BASE CABINETS		LF	165.00		
	ADDITIONAL BASE CABINETS AS REQUIRED EA = EACH CABINET					
DRAWER BASE	WITH 4 DRAWERS	15"	EA	372.00		
		18"	EA	378.00		
		21"	EA	384.00		
		24"	EA	390.00		
SINK OR BURNER BASE		30"	EA	276.00		
		36"	EA	285.00		
		42"	EA	309.00		
SINK OR BURNER FRONT	FRONT AND FLOOR	30"	EA	240.00		
	(**NO** SIDES)	36"	EA	261.00		
		42"	EA	276.00		
		48"	EA	285.00		
ISLAND BASE	WITH DOORS	24"	EA	414.00		
	2 SIDES AND	36"	EA	558.00		
	DRAWERS 1 SIDE	42"	EA	594.00		
		48"	EA	636.00		
CORNER BASE	(LAZY SUSAN) WITH 36" ALONG EACH WALL		EA	393.00		
BROOM CABINET	***Width*** / ***Height*** / ***Depth***					
	18" / 7-0 / 12"		EA	429.00		
	18" / 7-0 / 24"		EA	477.00		
OVEN CABINET	***Width*** / ***Height*** / ***Depth***					
	24" / 7-0 / 24"		EA	681.00		
	27" / 7-0 / 24"		EA	723.00		

21

SPECIFICATIONS			UNIT	PRICE	LOCAL AREA MODIFICATION %	LOCAL AREA MODIFICATION PRICE
WALL CABINETS	THE BASIC WALL CABINET INSTALLATION INCLUDES THE FOLLOWING: • TWO "ABOVE APPLIANCE" WALL CABINETS 15" HIGH AND 36" WIDE @ $115.00 EACH RETAIL • 1 AND 2 DOOR WALL CABINETS 30" HIGH AS REQUIRED LF = TOTAL OVERALL FRONT WIDTH OF ALL WALL CABINETS		LF	108.00		
	ADDITIONAL WALL CABINETS AS REQUIRED EA = EACH CABINET					
UPPER ISLAND	WITH DOORS TWO SIDES, 24" HIGH	36"	EA	396.00		
	SAME AS ABOVE, 30" HIGH	24"	EA	360.00		
		27"	EA	375.00		
		30"	EA	396.00		
		33"	EA	411.00		
		36"	EA	429.00		
		42"	EA	459.00		
		48"	EA	498.00		
CORNER WALL	(LAZY SUSAN) WITH 24" ALONG EACH WALL		EA	285.00		
MICROWAVE WALL	MICROWAVE WALL CABINET, 24"		EA	453.00		

21. KITCHEN CABINETS, PREMIUM

SPECIFICATIONS			UNIT	PRICE	LOCAL AREA MODIFICATION	
					%	PRICE
BASE CABINETS	INSTALL PREFINISHED OR UNFINISHED KITCHEN CABINETS ON PLUMB & STRAIGHT WALLS. STAINING AND PAINTING **NOT** INCLUDED BASIC BASE CABINET INSTALLATION INCLUDES THE FOLLOWING • ONE 36" WIDE, 2 DOOR AND 2 DRAWER FRONTS, SINK BASE CABINET @ $230.00 EACH RETAIL • ONE 24" WIDE, 4 DRAWER DRAWER BASE CABINET • 1 AND 2 DOOR, 1 AND 2 DRAWER BASE CABINETS AS REQUIRED LF = TOTAL OVERALL FRONT WIDTH OF ALL BASE CABINETS		LF	222.00		
	ADDITIONAL BASE CABINETS AS REQUIRED EA = EACH CABINET					
DRAWER BASE	WITH 4 DRAWERS	15"	EA	522.00		
		18"	EA	531.00		
		21"	EA	534.00		
		24"	EA	552.00		
SINK OR BURNER BASE		30"	EA	381.00		
		36"	EA	393.00		
		42"	EA	432.00		
SINK OR BURNER FRONT	FRONT AND FLOOR	30"	EA	330.00		
	(**NO** SIDES)	36"	EA	363.00		
		42"	EA	381.00		
		48"	EA	390.00		
ISLAND BASE	WITH DOORS	24"	EA	663.00		
	2 SIDES AND	36"	EA	726.00		
	DRAWERS 1 SIDE	42"	EA	786.00		
		48"	EA	849.00		
CORNER BASE	(LAZY SUSAN) WITH 36" ALONG EACH WALL		EA	537.00		
BROOM CABINET	***Width*** / ***Height*** / ***Depth***					
	18" / 7-0 / 12"		EA	618.00		
	18" / 7-0 / 24"		EA	672.00		
OVEN CABINET	***Width*** / ***Height*** / ***Depth***					
	24" / 7-0 / 24"		EA	891.00		
	27" / 7-0 / 24"		EA	942.00		

21

SPECIFICATIONS			UNIT	PRICE	LOCAL AREA MODIFICATION %	LOCAL AREA MODIFICATION PRICE
WALL CABINETS	THE BASIC WALL CABINET INSTALLATION INCLUDES THE FOLLOWING: • TWO "ABOVE APPLIANCE" WALL CABINETS 15" HIGH AND 36" WIDE @ $170.00 EACH RETAIL • 1 AND 2 DOOR WALL CABINETS 30" HIGH AS REQUIRED LF = TOTAL OVERALL FRONT WIDTH OF ALL WALL CABINETS		LF	144.00		
	ADDITIONAL WALL CABINETS AS REQUIRED EA = EACH CABINET					
UPPER ISLAND	WITH DOORS TWO SIDES, 24" HIGH	36"	EA	540.00		
	SAME AS ABOVE, 30" HIGH	24"	EA	483.00		
		27"	EA	504.00		
		30"	EA	525.00		
		33"	EA	549.00		
		36"	EA	582.00		
		42"	EA	627.00		
		48"	EA	669.00		
CORNER WALL	(LAZY SUSAN) WITH 24" ALONG EACH WALL		EA	387.00		
MICROWAVE WALL	MICROWAVE WALL CABINET, 24"		EA	639.00		

21

21. KITCHEN CABINETS, TOP QUALITY

SPECIFICATIONS			UNIT	PRICE	LOCAL AREA MODIFICATION %	LOCAL AREA MODIFICATION PRICE
BASE CABINETS	INSTALL PREFINISHED OR UNFINISHED KITCHEN CABINETS ON PLUMB & STRAIGHT WALLS. STAINING AND PAINTING **NOT** INCLUDED BASIC BASE CABINET INSTALLATION INCLUDES THE FOLLOWING • ONE 36" WIDE, 2 DOOR AND 2 DRAWER FRONTS, SINK BASE CABINET @ $300.00 EACH RETAIL • ONE 24" WIDE, 4 DRAWER DRAWER BASE CABINET • 1 AND 2 DOOR, 1 AND 2 DRAWER BASE CABINETS AS REQUIRED LF = TOTAL OVERALL FRONT WIDTH OF ALL BASE CABINETS		LF	279.00		
	ADDITIONAL BASE CABINETS AS REQUIRED EA = EACH CABINET					
DRAWER BASE	WITH 4 DRAWERS	15"	EA	651.00		
		18"	EA	672.00		
		21"	EA	678.00		
		24"	EA	702.00		
SINK OR BURNER BASE		30"	EA	477.00		
		36"	EA	492.00		
		42"	EA	543.00		
SINK OR BURNER FRONT	FRONT AND FLOOR	30"	EA	414.00		
	(**NO** SIDES)	36"	EA	459.00		
		42"	EA	477.00		
		48"	EA	492.00		
ISLAND BASE	WITH DOORS	24"	EA	846.00		
	2 SIDES AND	36"	EA	924.00		
	DRAWERS 1 SIDE	42"	EA	1005.00		
		48"	EA	1092.00		
CORNER BASE	(LAZY SUSAN) WITH 36" ALONG EACH WALL		EA	681.00		
BROOM CABINET	***Width*** / ***Height***	***Depth***				
	18" / 7-0	12"	EA	978.00		
	18" / 7-0	24"	EA	1041.00		
OVEN CABINET	***Width*** / ***Height***	***Depth***				
	24" / 7-0	24"	EA	1134.00		
	27" / 7-0	24"	EA	1197.00		

SPECIFICATIONS			UNIT	PRICE	LOCAL AREA MODIFICATION	
					%	PRICE
WALL CABINETS	THE BASIC WALL CABINET INSTALLATION INCLUDES THE FOLLOWING: • TWO "ABOVE APPLIANCE" WALL CABINETS 15" HIGH AND 36" WIDE @ $220.00 EACH RETAIL • 1 AND 2 DOOR WALL CABINETS 30" HIGH AS REQUIRED LF = TOTAL OVERALL FRONT WIDTH OF ALL WALL CABINETS		LF	180.00		
	ADDITIONAL WALL CABINETS AS REQUIRED EA = EACH CABINET					
UPPER ISLAND	WITH DOORS TWO SIDES, 24" HIGH	36"	EA	684.00		
	SAME AS ABOVE, 30" HIGH	24"	EA	606.00		
		27"	EA	636.00		
		30"	EA	663.00		
		33"	EA	699.00		
		36"	EA	738.00		
		42"	EA	795.00		
		48"	EA	849.00		
CORNER WALL	(LAZY SUSAN) WITH 24" ALONG EACH WALL		EA	480.00		
MICROWAVE WALL	MICROWAVE WALL CABINET,	24"	EA	789.00		

21. KITCHEN FIXTURES & APPLIANCES, NEW & REPLACEMENT

SPECIFICATIONS			UNIT	PRICE	LOCAL AREA MODIFICATION %	LOCAL AREA MODIFICATION PRICE
	THE PRICES BELOW INCLUDE ONLY THE FIXTURES, APPLIANCES AND FITTINGS, AND ARE FOR NEW INSTALLATIONS THAT ARE ALREADY ROUGHED IN, OR WHERE EXISTING FIXTURES AND APPLIANCES HAVE BEEN REMOVED. THE LABOR INCLUDES ONLY THE COST OF SETTING IN PLACE THE FIXTURES AND APPLIANCES BY CARPENTERS ON THE JOB. FOR PLUMBING AND ELECTRICAL ROUGH IN AND INSTALLATION COSTS, SEE SECTIONS 14 AND 16.					
REFRIGER-ATOR	REFRIGERATOR	ECONOMY	EA	594.00		
		BUILDER	EA	1050.00		
		PREMIUM	EA	1590.00		
CLOTHES DRYER	ELECTRIC CLOTHES DRYER	ECONOMY	EA	474.00		
		BUILDER	EA	669.00		
		PREMIUM	EA	903.00		
ELECTRIC COOKTOP, OVEN AND RANGE	ELECTRIC COOKTOP WITH DUCTED DOWNDRAFT	35" X 20"	EA	435.00		
	WALL OVEN	SINGLE	EA	393.00		
		DOUBLE	EA	477.00		
	FREE-STANDING ELECTRIC RANGE AND OVEN	ECONOMY	EA	555.00		
		BUILDER	EA	720.00		
		PREMIUM	EA	1140.00		
MICROWAVE	MICROWAVE OVEN	ECONOMY	EA	294.00		
		BUILDER	EA	330.00		
		PREMIUM	EA	405.00		
TRASH COMPACTOR	UNDER-COUNTER COMPACTOR	ECONOMY	EA	594.00		
		BUILDER	EA	735.00		
		PREMIUM	EA	864.00		
	FOR INSTANT HOT, SEE PAGE 206					

21

SPECIFICATIONS		UNIT	PRICE	LOCAL AREA MODIFICATION %	LOCAL AREA MODIFICATION PRICE
BULKHEAD, DROP, SOFFIT	DRYWALL, FROM CEILING TO TOP OF WALL CABINET, INCLUDING FRAMING AS REQUIRED	LF	16.95		
COUNTER-TOP	WITH PLASTIC, METAL OR SELF EDGE AND 4" BACKSPLASH, SHOP-BUILT BY OTHERS				
	ECONOMY	LF	22.11		
	BUILDER	LF	33.36		
	PREMIUM	LF	56.61		
	SOLID SURFACE COUNTERTOP, 4" BACKSPLASH				
	1/2"	LF	150.00		
	3/4"	LF	213.00		
CUTTING BOARD	1-1/2" X 18" X 25" MAPLE LAMINATED, INSTALLED IN COUNTERTOP ABOVE	EA	240.00		
CERAMIC TILE	CERAMIC TILE COUNTERTOP 4-1/4 " X 4-1/4", 6" X 6", AND 8" X 8", INCLUDES TRIM, SET IN MUD	SF	18.65		
	SAME AS ABOVE, SET IN THIN SET OR MASTIC	SF	12.42		
KITCHEN CABINET REFACING	REFACE EXISTING KITCHEN CABINETS WITH PLASTIC LAMINATE DONE IN PLACE, INCLUDING NEW DOOR HARDWARE				
	WALL CABINET DOOR	EA	57.00		
	BASE CABINET DOOR	EA	75.00		
	WALL CABINET FACE	EA	39.00		
	BASE CABINET FACE	EA	48.00		
	REFACE SIDE OF CABINET WITH PLASTIC LAMINATE				
	WALL CABINET	EA	36.00		
	BASE CABINET	EA	48.00		

21

22. BATHROOM CABINETS & ACCESSORIES

SPECIFICATIONS		UNIT	PRICE	LOCAL AREA MODIFICATION	
				%	PRICE
MEDICINE CABINET	RECESSED CABINET WITH HINGED WINDOW GLASS MIRROR DOOR, OVERALL SIZE 16" X 22"	EA	75.00		
	SAME AS ABOVE WITH PLATE GLASS MIRROR DOOR	EA	96.00		
	RECESSED CABINET WITH HINGED PLATE GLASS MIRROR DOOR, 2 FLUORESCENT SIDELIGHTS, OVERALL SIZE 22" X 22"	EA	132.00		
	RECESSED CABINET WITH OVERHEAD FLUORES-CENT FIXTURE, 2 MIRROR DOORS, OVERALL SIZE 27" X 23"	EA	117.00		
SURFACE MOUNT CABINET	SURFACE MOUNT CABINET WITH 2 SLIDING MIR-ROR DOORS, OVERALL SIZE 24" X 16"	EA	93.00		
	SAME AS ABOVE WITH OVERHEAD 4-BULB FIX-TURE				
	Overall Size				
	24" X 23"	EA	129.00		
	28" X 23"	EA	132.00		
	36" X 23"	EA	153.00		
MIRROR	WALL MIRROR 1/4" THICK WITH POLISHED EDGES, ATTACHED TO WALL WITH CLIPS OR ADHESIVE	SF	14.70		
CHROME ACCES-SORIES	POLISHED CHROME BATH ACCESSORIES: TOWEL BAR, TOOTHBRUSH AND GLASS HOLDER, PAPER HOLDER, SOAP DISH	SET	96.00		
SHOWER ROD	CHROME SHOWER CURTAIN ROD, 5'	EA	60.00		
GRAB BAR	STAINLESS STEEL GRAB BAR, 1-1/4" DIAMETER, STRAIGHT, WITH ANCHOR PLATES				
	24"	EA	60.00		
	36"	EA	63.00		
	NOTE: FOR CERAMIC BATH ACCESSORIES, SEE P. 230				

22

SPECIFICATIONS		UNIT	PRICE	LOCAL AREA MODIFICATION	
				%	PRICE
FOLDING DOORS	TUB ENCLOSURE, ALUMINUM FRAME, PLASTIC PANELS, ROLLER GLIDES TOP AND BOTTOM, FOLD TO 12" FOR UP TO 66" OPENING, 58" HIGH	EA	246.00		
	SHOWER ENCLOSURE, SAME SPECIFICATIONS AS ABOVE, FOR OPENING 30" TO 66" WIDE AND 68" HIGH	EA	225.00		
BY-PASS DOOR	TUB ENCLOSURE, ALUMINUM FRAME, 2 PLASTIC PANELS 58" HIGH FOR UP TO 60" OPENING	EA	213.00		
	SAME AS ABOVE WITH TEMPERED GLASS	EA	267.00		
	SHOWER ENCLOSURE WITH 2 TEMPERED GLASS DOORS 70" HIGH FOR UP TO 48" OPENING	EA	312.00		
HINGED DOOR	ALUMINUM FRAME, ONE HINGED TEMPERED GLASS DOOR 64" HIGH FOR 23" TO 28" OPENINGS	EA	258.00		
	SAME AS ABOVE WITH ONE STATIONARY PANEL AND ONE HINGED DOOR FOR 48" OPENING	EA	387.00		
NEO-ANGLE GLASS SHOWER ENCLOSURE	ALUMINUM FRAME, 24" SIDE PANELS, 27" DOOR, 72" HIGH, OBSCURE GLASS, INSTALLED	EA	363.00		
	SAME AS ABOVE, WITH BRIGHT BRASS FINISH	EA	423.00		
TUB WALL KIT	TUB SURROUND 3, 4 OR 5 PIECE VINYL INSTEAD OF CERAMIC TILE, **NO** DOOR INCLUDED				
	ECONOMY	EA	162.00		
	BUILDER	EA	267.00		
	PREMIUM	EA	492.00		

22. SHUTTERS, METAL FIREPLACE

SPECIFICATIONS			UNIT	PRICE	LOCAL AREA MODIFICATION %	LOCAL AREA MODIFICATION PRICE
DOOR SHUTTERS	EXTERIOR DOOR SHUTTERS WITH STATIONARY SLATS, INSTALLED WITHOUT HINGES					
	WOOD		PAIR	105.00		
	PLASTIC		PAIR	105.00		
	ALUMINUM		PAIR	114.00		
WINDOW SHUTTERS	EXTERIOR WINDOW SHUTTERS WITH STATIONARY SLATS, INSTALLED WITHOUT HINGES					
	WOOD		PAIR	78.00		
	PLASTIC		PAIR	78.00		
	ALUMINUM		PAIR	90.00		
INTERIOR SHUTTERS	INTERIOR MOVABLE SHUTTER INSTALLED (HINGED), FOUR PANELS TO EACH SET, 3/4" PINE, PRE-ASSEMBLED WITH KNOBS, HINGES, GLUED JOINTS					
	Width	***Height***				
	25" TO 29"	17" TO 21"	SET	78.00		
		33" TO 37"	SET	96.00		
	29" TO 33"	17" TO 21"	SET	99.00		
		33" TO 37"	SET	120.00		
	33" TO 37"	17" TO 21"	SET	105.00		
		33" TO 37"	SET	120.00		
	37" TO 41"	17" TO 21"	SET	114.00		
		33" TO 37"	SET	132.00		
ZERO CLEARANCE FIREPLACE	PRE-FABRICATED METAL WOOD BURNING FIRE-PLACE AND CHIMNEY, INCLUDING EXTERIOR CHIM-NEY HOUSING, 4-5 X 5-0, 13 FEET MEASURING FROM FLOOR TO TOP OF CHIMNEY HOUSING		EA	1617.00		
	ADDITIONAL CHIMNEY ABOVE 13 FEET		LF	37.50		
	COLORED FIREPLACE AND CHIMNEY	**ADD**	EA	105.00		

22

SPECIFICATIONS			UNIT	PRICE	LOCAL AREA MODIFICATION %	LOCAL AREA MODIFICATION PRICE
BURGLAR ALARM	CONTROL BOX					
		STANDARD	EA	1005.00		
		PREMIUM	EA	1545.00		
	SENSOR	HARD WIRED (WIRES IN WALLS)	EA	72.00		
		EXPOSED WIRES	EA	72.00		
		RADIO TRANSMITTER	EA	132.00		
ACCES-SORIES	OUTSIDE HORN		EA	210.00		
	EMERGENCY POCKET TRANSMITTER		EA	123.00		
	AUTOMATIC MESSAGE TRANSMITTER		EA	765.00		
	FREEZE ALERT		EA	135.00		
	MOTION DETECTOR		EA	450.00		
	REMOTE ALARM		EA	120.00		
FIRE EXTIN-GUISHER	INSTALL		EA	75.00		
	RE-TEST		EA	30.00		
SECURITY BARS	SECURITY BARS INSTALLED ON EXISTING WINDOW		SF	18.00		
	SF = WINDOW					
	HINGED DOOR BARS WITH DEADBOLT		EA	294.00		

22. WOOD FENCING

SPECIFICATIONS		UNIT	PRICE	LOCAL AREA MODIFICATION %	LOCAL AREA MODIFICATION PRICE
	WOOD FENCING COSTS BELOW INCLUDE 4" X 4" POSTS 8'-0" O.C., SET 2" INTO 6" GRAVEL BED AT BOTTOM OF 36" DEEP POST HOLE AND TAMPED DIRT AND GRAVEL TO TOP OF POST HOLE.				
	IF, INSTEAD OF TAMPED DIRT AND GRAVEL, POSTS ARE SET IN GRAVEL BED WITH DRY CONCRETE PLACED ON GRAVEL BED TO TOP OR NEAR TOP OF POST HOLE **ADD**	LF	.90		
	LF = FENCING				
BASKET-WEAVE	3/4" X 6" REDWOOD, INCLUDING 4" X 4" LINE POSTS, CORNER AND END POSTS				
	Height				
	48"	LF	20.15		
	60"	LF	26.96		
	SAME AS ABOVE, PLUS 12" CRISS-CROSS ALONG TOP				
	48"	LF	25.46		
	60"	LF	30.87		
	2 GATEPOSTS AND 42" WIDE GATE				
	48"	EA	165.00		
	60"	EA	201.00		
PICKET	1" X 4" CEDAR OR REDWOOD PICKETS 6" O.C., 4" X 4" POSTS, 2-RAIL FENCE				
	36"	LF	13.17		
	48"	LF	14.30		
	60"	LF	16.47		
	2 GATEPOSTS AND 42" WIDE GATE				
	36"	EA	144.00		
	48"	EA	150.00		
	60"	EA	159.00		
STOCKADE	1" X 3" CEDAR PICKETS WITH 4" X 4" OR 6" ROUND POSTS, (3) 2" X 3" RAILS72"	LF	21.17		
	2 GATEPOSTS AND 42" WIDE GATE, 72"	EA	234.00		
PRIVACY	PRESSURE TREATED 1 X 6, (3) 2 X 4 RAILS, 4 X 4 POSTS, 8' OC, 72"	LF	14.70		
	2 GATEPOSTS AND 36" WIDE GATE, 72"	EA	192.00		

22

SPECIFICATIONS		UNIT	PRICE	LOCAL AREA MODIFICATION %	LOCAL AREA MODIFICATION PRICE
SHADOW BOX	OR "ALTERNATING BOARD" PRESSURE TREATED 1 X 6, (3) 2 X 4 RAILS, 4 X 4 POSTS, 8' OC				
	72"	LF	17.27		
	2 GATEPOSTS AND 36" WIDE GATE, 72"	EA	228.00		
SPLIT RAIL	SPLIT RAIL FENCE WITH 10 FEET BETWEEN LINE POSTS				
	Height				
	2 RAIL 36"	LF	9.69		
	3 RAIL 42"	LF	11.55		
RANCH ROUND (DUXBURY)	INCLUDING FULL ROUND PINE RAILS AND POSTS, 8 FEET BETWEEN LINE POSTS				
	36"	LF	9.15		
	42"	LF	10.73		
	2 GATEPOSTS AND 42" WIDE GATE				
	36"	EA	114.00		
CHAIN LINK	11 GA., GALVANIZED STEEL OR 11-1/2 GA., VINYL COVERED FENCING				
	36"	LF	8.24		
	42"	LF	9.66		
	48"	LF	10.13		
	60"	LF	11.60		
	72"	LF	12.60		
	2 GATEPOSTS AND 42" WIDE GATE				
	36"	EA	174.00		
	42"	EA	180.00		
	48"	EA	186.00		
	60"	EA	201.00		
	72"	EA	210.00		
	2 GATEPOSTS AND 12 FOOT WIDE (DOUBLE) GATE				
	36"	EA	294.00		
	42"	EA	297.00		
	48"	EA	318.00		
	60"	EA	345.00		
	72"	EA	372.00		

22

23. HARDWOOD FLOORS

SPECIFICATIONS		UNIT	PRICE	LOCAL AREA MODIFICATION %	LOCAL AREA MODIFICATION PRICE
HARDWOOD	LAY FLOOR WITH T&G AND END-MATCHED OAK FLOORING, 25/32" X 2-1/4"				
	CLEAR	SF	9.84		
	SELECT	SF	8.10		
	#1 COMMON	SF	7.17		
PRE-FINISHED HARDWOOD	PREFINISHED OAK, 25/32 X 2-1/4	SF	10.76		
	RANCH PLANK WITH SCREWS AND PLUGS	SF	14.07		
HARDWOOD TOOTHED IN	STRIP OAK FLOORING TOOTHED IN TO EXISTING FLOOR, 25/32" X 2-1/4" **ADD**	EA	24.00		
	EA = EACH TOOTH-IN JOINT				
PARQUET	OAK LAMINATED, SET IN MASTIC OVER CONCRETE OR PLYWOOD DECK, T&G, PREFINISHED, CLEAR				
	12" X 12" X 1/2"	SF	9.05		
FLOOR FINISHING	SAND AND FINISH NEW FLOOR				
	NATURAL	SF	1.77		
	DARK STAIN	SF	2.78		
	BLEACHED	SF	3.26		
	SAND AND FINISH OLD FLOOR				
	NATURAL	SF	2.16		
	DARK STAIN	SF	3.41		
	BLEACHED	SF	3.89		

SPECIFICATIONS		UNIT	PRICE	LOCAL AREA MODIFICATION	
				%	PRICE
UNDER-LAYMENT	UNDERLAYMENT FOR RESILIENT FLOORS NAILED OVER EXISTING FLOOR				
	PLYWOOD 3/8"	SF	1.29		
	1/2"	SF	1.41		
	5/8"	SF	1.59		
	3/4"	SF	1.74		
	PARTICLEBOARD 3/8"	SF	1.02		
	1/2"	SF	1.07		
	5/8"	SF	1.11		
	3/4"	SF	1.25		
ASPHALT TILE	LAID IN ADHESIVE OVER SMOOTH SURFACE, 9" X 9"				
	B GRADE	SF	2.21		
	C GRADE	SF	2.45		
	D GRADE	SF	2.76		
VINYL COM-POSITION	LAID IN ADHESIVE OVER SMOOTH SURFACE, 9" X 9"				
	1/16"	SF	1.82		
	3/32"	SF	2.19		
	1/8"	SF	3.68		
	SAME AS ABOVE, 12" X 12"				
	1/16"	SF	1.68		
	3/32"	SF	2.06		
	1/8"	SF	3.53		
VINYL TILE	LAID IN ADHESIVE OVER SMOOTH SURFACE, 9" X 9"				
	1/16"	SF	2.64		
	3/32"	SF	3.84		
	1/8"	SF	5.34		
	SAME AS ABOVE, 12" X 12"				
	1/16"	SF	2.55		
	1/8"	SF	5.25		
	PREMIUM	SF	6.00		
SHEET VINYL	LAID IN ADHESIVE OVER SMOOTH SURFACE				
	ECONOMY	SF	2.55		
	MEDIUM	SF	3.81		
	PREMIUM	SF	6.36		
BASE	VINYL BASE, INCLUDING CORNERS				
	2-1/4"	LF	1.58		
	4"	LF	1.68		
	6"	LF	1.85		

23. TILE FLOORS

SPECIFICATIONS		UNIT	PRICE	LOCAL AREA MODIFICATION	
				%	PRICE
UNDER-LAYMENT	WATER RESISTANT 1/2" THICK BACKER BOARD UNDERLAYMENT FOR CERAMIC TILE FLOOR, INCLUDING MORTAR, JOINT TAPE AND NAILS	SF	4.14		
CERAMIC TILE	CERAMIC TILE FLOOR, 4-1/4" X 4-1/4", 6" X 6" AND 8" X 8", INCLUDING COVE, BASE OR BULLNOSE TRIM, SET IN MUD	SF	19.14		
	SAME AS ABOVE, SET IN THIN SET OR MASTIC	SF	12.14		
QUARRY TILE	QUARRY TILE 1/2" OR 3/4" THICK, 4" X 4", 6" X 6", 8" X 8" AND 9" X 9"	SF	12.14		
PAVER TILE	PAVER TILE 1" THICK, 4" X 4", 6" X 6"	SF	13.79		
MARBLE	MARBLE TILE 3/8" THICK, 12" X 12"	SF	22.53		
SLATE	SLATE TILE, 3/8" THICK, 12" X 12"	SF	15.12		

23

SPECIFICATIONS		UNIT	PRICE	LOCAL AREA MODIFICATION	
				%	PRICE
CARPETING	INDOOR-OUTDOOR CARPETING ON WOOD OR CONCRETE				
	ECONOMY	SF	1.28		
	MEDIUM	SF	1.65		
	PREMIUM	SF	2.06		
	INTERIOR CARPETING, INCLUDING PADDING				
	@ $12/YD	SF	3.00		
	@ $15/YD	SF	3.77		
	@ $20/YD	SF	4.89		
	@ $25/YD	SF	6.03		
	STAIR CARPETING **ADD**	Per Riser	7.50		
REMOVE CARPET	REMOVE EXISTING CARPETING FROM FLOOR				
	BONDED	SF	.38		
	TACKLESS	SF	.11		
	CARPET PAD	SF	.08		
CLEAN	CLEAN EXISTING CARPETING (IN PLACE) WITH STEAM	SF	.60		

24. PAINTING PREPARATION, EXTERIOR

SPECIFICATIONS		UNIT	PRICE	LOCAL AREA MODIFICATION	
				%	PRICE
PAINT REMOVAL	REMOVE EXISTING PAINT OR VARNISH WITH CHEMICALS, 1 OR 2 LAYERS PER COAT OF PAINT REMOVER				
	EXTERIOR WOOD SIDING	SF	1.56		
	EXTERIOR TRIM	LF	1.34		
	REMOVE NON LEAD BASED PAINT BY LIGHT SAND BLASTING	SF	1.38		
	REMOVE LOOSE OR BLISTERED PAINT WITH WIRE BRUSH OR PAINT SCRAPER	SF	.20		
BURNING OFF	EXTERIOR SIDING, NON LEAD BASED PAINT				
	AVERAGE	SF	1.49		
	EXTENSIVE	SF	2.06		
	EXTERIOR TRIM, NON LEAD BASED PAINT	LF	1.13		
SANDING AND FILLING	SAND AND FILL EXTERIOR WOOD SIDING AND TRIM				
	AVERAGE	SF	.24		
	EXTENSIVE	SF	.30		
	SAND AND FILL EXTERIOR TRIM ONLY				
	AVERAGE	LF	.30		
	EXTENSIVE	LF	.41		
REGLAZE	REGLAZE SASH	LF	1.13		
ACID WASH	ACID WASH GALVANIZED GUTTER AND DOWN-SPOUT	LF	.41		
CAULK	CAULK TRIM OR SIDING JOINTS	LF	.63		
WASHING	POWER WASH WITH TSP IN PREPARATION FOR EXTERIOR PAINTING	SF	.20		
	EACH ADDITIONAL STORY IF LADDER OR SCAFFOLDING IS REQUIRED **ADD**	SF	20%		

SPECIFICATIONS		UNIT	PRICE	LOCAL AREA MODIFICATION %	LOCAL AREA MODIFICATION PRICE
	NOTE: PAINTING PRICES ON EVERY PAGE INCLUDE SET-UP AND CLEAN-UP TIME AND PROTECTION OF ITEMS NOT TO BE PAINTED. ALL WORK SHOWN IS AT ONE-STORY LEVEL EXCEPT AS NOTED.				
SIDING	WOOD, PLYWOOD, ASBESTOS AND COMPOSITION SIDING, PAINT WITH BRUSH				
	1 COAT ON PAINTED SURFACE	SF	.50		
	PRIME AND 1 COAT	SF	.87		
	PRIME AND 2 COATS	SF	1.25		
	SAME AS ABOVE, WITH ROLLER				
	1 COAT ON PAINTED SURFACE	SF	.44		
	PRIME AND 1 COAT	SF	.68		
	PRIME AND 2 COATS	SF	.95		
	SAME AS ABOVE, WITH SPRAY GUN				
	1 COAT ON PAINTED SURFACE	SF	.26		
	PRIME AND 1 COAT	SF	.48		
	PRIME AND 2 COATS	SF	.72		
	SAME AS ABOVE, STAINED, WITH BRUSH				
	1 COAT	SF	.45		
	2 COATS	SF	.80		
ALUMINUM SIDING	ALUMINUM SIDING, PAINT WITH BRUSH				
	1 COAT	SF	.32		
	2 COATS	SF	.50		
	SAME AS ABOVE, WITH SPRAY GUN				
	1 COAT	SF	.26		
	2 COATS	SF	.48		
	FOR EACH ADDITIONAL STORY UP, PER COAT **ADD**	SF	20%		

24. SHINGLES, STUCCO, MASONRY PAINTING

SPECIFICATIONS		UNIT	PRICE	LOCAL AREA MODIFICATION %	LOCAL AREA MODIFICATION PRICE
WOOD SHINGLES OR ROUGH SIDING	CEDAR SHINGLES, PAINT WITH BRUSH				
	1 COAT ON PAINTED SURFACE	SF	.59		
	PRIME AND 1 COAT	SF	.99		
	PRIME AND 2 COATS	SF	1.43		
	CEDAR SHINGLES, WITH SPRAY GUN				
	1 COAT ON PAINTED SURFACE	SF	.29		
	PRIME AND 1 COAT	SF	.47		
	PRIME AND 2 COATS	SF	.69		
	CEDAR SHINGLES, STAINED WITH BRUSH				
	1 COAT	SF	.56		
	2 COATS	SF	.93		
	CEDAR SHINGLES, STAINED WITH SPRAY GUN				
	1 COAT	SF	.24		
	2 COATS	SF	.45		
SMOOTH BRICK OR STUCCO	WITH ROLLER				
	1 COAT ON PAINTED SURFACE	SF	.54		
	PRIME AND 1 COAT	SF	.87		
	PRIME AND 2 COATS	SF	1.20		
	WITH SPRAY GUN				
	1 COAT ON PAINTED SURFACE	SF	.26		
	PRIME AND 1 COAT	SF	.47		
	PRIME AND 2 COATS	SF	.71		
POROUS BRICK OR STUCCO	WITH ROLLER				
	1 COAT ON PAINTED SURFACE	SF	.66		
	PRIME AND 1 COAT	SF	1.13		
	PRIME AND 2 COATS	SF	1.55		
	WITH SPRAY GUN				
	1 COAT ON PAINTED SURFACE	SF	.27		
	PRIME AND 1 COAT	SF	.51		
	PRIME AND 2 COATS	SF	.78		

SPECIFICATIONS		UNIT	PRICE	LOCAL AREA MODIFICATION %	LOCAL AREA MODIFICATION PRICE
MOULDING AND TRIM	DOOR OR WINDOW MOULDING OR TRIM (IF SAME COLOR AND FINISH AS WALL, INCLUDE IN WALL ESTIMATE), WITH BRUSH				
	1 COAT ON PAINTED SURFACE	LF	.41		
	PRIME AND 1 COAT	LF	.72		
	PRIME AND 2 COATS	LF	1.05		
FASCIA OR RAKE	UP TO 12" FASCIA OR RAKE WITH BRUSH, ONE FACE & EDGE				
	1 COAT ON PAINTED SURFACE	LF	.57		
	PRIME AND 1 COAT	LF	1.02		
	PRIME AND 2 COATS	LF	1.29		
SOFFIT	UP TO 12" SOFFIT, WITH BRUSH				
	1 COAT ON PAINTED SURFACE	LF	.69		
	PRIME AND 1 COAT	LF	1.28		
	PRIME AND 2 COATS	LF	1.86		
	SOFFIT, WITH ROLLER				
	1 COAT ON PAINTED SURFACE	SF	.47		
	PRIME AND 1 COAT	SF	.80		
	PRIME AND 2 COATS	SF	1.07		
	SOFFIT, WITH SPRAY GUN				
	1 COAT ON PAINTED SURFACE	SF	.26		
	PRIME AND 1 COAT	SF	.47		
	PRIME AND 2 COATS	SF	.69		
EAVES, OPEN WITH RAFTERS	WITH BRUSH, ALL SIDES OF RAFTERS, INTERIOR OF FASCIA AND UNDERSIDE OF ROOF SHEATHING				
	1 COAT ON PAINTED SURFACE	SF	.74		
	PRIME AND 1 COAT	SF	1.46		
	PRIME AND 2 COATS	SF	2.19		
	WITH SPRAY GUN				
	1 COAT ON PAINTED SURFACE	SF	.29		
	PRIME AND 1 COAT	SF	.53		
	PRIME AND 2 COATS	SF	.75		

24. DECK, PORCH PAINTING

SPECIFICATIONS		UNIT	PRICE	LOCAL AREA MODIFICATION	
				%	PRICE
WOOD DECK OR PORCH FLOOR, TOP SIDE	PAINT WITH BRUSH				
	1 COAT ON PAINTED SURFACE	SF	.51		
	PRIME AND 1 COAT	SF	.78		
	PRIME AND 2 COATS	SF	1.05		
	PAINT WITH ROLLER				
	1 COAT ON PAINTED SURFACE	SF	.51		
	PRIME AND 1 COAT	SF	.75		
	PRIME AND 2 COATS	SF	.98		
	PAINT WITH SPRAY GUN				
	1 COAT ON PAINTED SURFACE	SF	.33		
	PRIME AND 1 COAT	SF	.54		
	PRIME AND 2 COATS	SF	.72		
	STAIN AND/OR CLEAR FINISH WITH BRUSH				
	1 COAT	SF	.50		
	2 COATS	SF	.84		
	3 COATS	SF	1.17		
	STAIN AND/OR CLEAR FINISH WITH ROLLER				
	1 COAT	SF	.47		
	2 COATS	SF	.75		
	3 COATS	SF	1.05		
	STAIN AND/OR CLEAR FINISH WITH SPRAY GUN				
	1 COAT	SF	.29		
	2 COATS	SF	.54		
	3 COATS	SF	.78		
PORCH CEILING	PAINT WOOD PORCH CEILING WITH BRUSH				
	1 COAT ON PAINTED SURFACE	SF	.50		
	PRIME AND 1 COAT	SF	.83		
	PRIME AND 2 COATS	SF	1.13		
	PAINT WITH ROLLER				
	1 COAT ON PAINTED SURFACE	SF	.47		
	PRIME AND 1 COAT	SF	.80		
	PRIME AND 2 COATS	SF	1.07		
	PAINT WITH SPRAY GUN				
	1 COAT ON PAINTED SURFACE	SF	.27		
	PRIME AND 1 COAT	SF	.51		
	PRIME AND 2 COATS	SF	.77		
CONCRETE PORCH	PAINT TOP SURFACE AND SIDES OF CONCRETE PORCH SLAB WITH ROLLER				
	1 COAT	SF	.66		
	2 COATS	SF	1.13		
	PAINT WITH SPRAY GUN				
	1 COAT	SF	.30		
	2 COATS	SF	.50		

	SPECIFICATIONS	UNIT	PRICE	LOCAL AREA MODIFICATION %	LOCAL AREA MODIFICATION PRICE
PORCH OR DECK RAILING	SF = HEIGHT OF RAILING MULTIPLIED BY LENGTH OF RAILING, **ONE SIDE ONLY**				
	COSTS SHOWN INCLUDE PAINTING ALL SIDES OF RAILINGS AND BALUSTERS,				
	WITH BRUSH 1 COAT	SF	1.11		
	2 COATS	SF	1.95		
	PAINT WITH SPRAY GUN				
	1 COAT	SF	.59		
	2 COATS	SF	1.13		
	STAIN AND/OR CLEAR FINISH WITH BRUSH				
	1 COAT	SF	1.13		
	2 COATS	SF	1.95		
	3 COATS	SF	2.76		
	STAIN AND/OR CLEAR FINISH WITH SPRAY GUN				
	1 COAT	SF	.51		
	2 COATS	SF	1.01		
	3 COATS	SF	1.52		
WOOD STEPS	SF = TO OBTAIN SF AREA, *MEASURE TOP SIDE ONLY* OF STEPS				
	PAINT ALL SURFACES (TOP, BOTTOM, SIDES) OF STEPS AND STRINGERS, PAINT WITH BRUSH				
	1 COAT ON PAINTED SURFACE	SF	1.05		
	PRIME AND 1 COAT	SF	1.74		
	PRIME AND 2 COATS	SF	2.60		
	PAINT WITH SPRAY GUN				
	1 COAT ON PAINTED SURFACE	SF	.50		
	PRIME AND 1 COAT	SF	.95		
	PRIME AND 2 COATS	SF	1.44		
	STAIN AND/OR CLEAR FINISH WITH BRUSH				
	1 COAT	SF	1.13		
	2 COATS	SF	1.95		
	3 COATS	SF	2.76		
	STAIN AND/OR CLEAR FINISH WITH SPRAY GUN				
	1 COAT	SF	.51		
	2 COATS	SF	1.01		
	3 COATS	SF	1.52		

24. EXTERIOR, FENCE PAINTING

SPECIFICATIONS		UNIT	PRICE	LOCAL AREA MODIFICATION	
				%	PRICE
WINDOW SCREEN	PAINT WOOD WINDOW SCREEN, ALL SIDES				
	1 COAT	EA	6.98		
	2 COATS	EA	12.08		
STORM SASH	PAINT STORM SASH ALL SIDES				
	1 COAT	EA	9.00		
	2 COATS	EA	15.23		
BLINDS AND SHUTTERS	WOOD OR METAL, LOUVERED, BOTH SIDES AND ALL EDGES, WITH BRUSH				
	1 COAT ON PAINTED SURFACE	EA	21.75		
	PRIME AND 1 COAT	EA	35.25		
	PRIME AND 2 COATS	EA	48.00		
	WITH SPRAY GUN				
	1 COAT ON PAINTED SURFACE	EA	10.50		
	PRIME AND 1 COAT	EA	16.50		
	PRIME AND 2 COATS	EA	21.00		
FENCE, PLAIN	PLAIN BOARD FENCE, ALL SIDES				
	BRUSH 1 COAT	SF	.99		
	2 COATS	SF	1.47		
	SPRAY 1 COAT	SF	.35		
	2 COATS	SF	.56		
	SF = HEIGHT OF FENCE MULTIPLIED BY LENGTH OF FENCE, **ONE FACE ONLY**				
PICKET	PICKET FENCE, ALL SIDES				
	BRUSH 1 COAT	SF	1.19		
	2 COATS	SF	2.06		
	SPRAY 1 COAT	SF	.23		
	2 COATS	SF	.41		
CHAIN LINK	CHAIN LINK FENCE, ALL SIDES				
	BRUSH 1 COAT	SF	.92		
	2 COATS	SF	1.58		
	ROLLER 1 COAT	SF	.45		
	2 COATS	SF	.78		
	CHAIN LINK FENCE, ALL SIDES WITH SPRAY GUN				
	1 COAT	SF	.29		
	2 COATS	SF	.44		

24

SPECIFICATIONS		UNIT	PRICE	LOCAL AREA MODIFICATION %	LOCAL AREA MODIFICATION PRICE
GUTTER AND/OR DOWNSPOUT	METAL GUTTER AND/OR DOWNSPOUT, PAINT WITH BRUSH				
	1 COAT ON PAINTED SURFACE	LF	.57		
	PRIME AND 1 COAT	LF	.81		
	PRIME AND 2 COATS	LF	1.07		
	WOOD GUTTER, 4" X 5", WITH BRUSH				
	1 COAT ON PAINTED SURFACE	LF	.66		
	PRIME AND 1 COAT	LF	1.08		
	PRIME AND 2 COATS	LF	1.41		
LATTICE WORK	PAINT ALL SIDES OF LATTICE WORK WITH BRUSH				
	SF = HEIGHT OF LATTICE WORK MULTIPLIED BY LENGTH, **ONE FACE ONLY**				
	1 COAT	SF	.84		
	2 COATS	SF	1.46		
	SAME AS ABOVE, WITH SPRAY GUN				
	1 COAT	SF	.26		
	2 COATS	SF	.44		
COLUMNS AND PILASTERS	WOOD COLUMNS AND PILASTERS, PAINT WITH BRUSH				
	1 COAT	SF	.48		
	2 COATS	SF	.83		
	WOOD COLUMNS AND PILASTERS, PAINT WITH SPRAY GUN 1 COAT	SF	.26		
	2 COATS	SF	.48		

24. DOOR PAINTING

SPECIFICATIONS		UNIT	PRICE	LOCAL AREA MODIFICATION	
				%	PRICE
FLUSH ENTRANCE DOOR	PAINT ONE SIDE ONLY, INCLUDING JAMB AND CASING, WITH BRUSH				
	1 COAT ON PAINTED SURFACE	EA	19.50		
	PRIME AND 1 COAT	EA	32.25		
	PRIME AND 2 COATS	EA	45.00		
	WITH ROLLER				
	1 COAT ON PAINTED SURFACE	EA	11.40		
	PRIME AND 1 COAT	EA	18.60		
	PRIME AND 2 COATS	EA	25.50		
	STAIN AND/OR CLEAR FINISH WITH BRUSH				
	1 COAT	EA	24.75		
	2 COATS	EA	41.70		
	3 COATS	EA	58.50		
2 TO 6 PANEL ENTRANCE DOOR	PAINT ONE SIDE ONLY, INCLUDING JAMB & CASINGS, WITH BRUSH				
	1 COAT ON PAINTED SURFACE	EA	24.00		
	PRIME AND 1 COAT	EA	40.50		
	PRIME AND 2 COATS	EA	54.75		
	STAIN AND/OR CLEAR FINISH WITH BRUSH				
	1 COAT	EA	31.05		
	2 COATS	EA	52.20		
	3 COATS	EA	70.05		

SPECIFICATIONS		UNIT	PRICE	LOCAL AREA MODIFICATION %	LOCAL AREA MODIFICATION PRICE
2 TO 6 LIGHT DOOR	PAINT ONE SIDE ONLY, INCLUDING JAMB AND CASING, WITH BRUSH				
	1 COAT ON PAINTED SURFACE	EA	29.70		
	PRIME AND 1 COAT	EA	46.50		
	PRIME AND 2 COATS	EA	63.45		
	STAIN AND/OR CLEAR FINISH, WITH BRUSH				
	1 COAT	EA	37.50		
	2 COATS	EA	60.00		
	3 COATS	EA	81.75		
9 TO 15 LIGHT FRENCH DOOR	ONE SIDE ONLY, INCLUDING JAMB AND CASING, WITH BRUSH				
	1 COAT ON PAINTED SURFACE	EA	34.20		
	PRIME AND 1 COAT	EA	56.55		
	PRIME AND 2 COATS	EA	78.00		
	STAIN AND/OR CLEAR FINISH, WITH BRUSH				
	1 COAT	EA	44.25		
	2 COATS	EA	75.00		
	3 COATS	EA	100.50		
COLONIAL ENTRANCE	FOR SINGLE DOOR, PAINT WITH BRUSH				
	1 COAT ON PAINTED SURFACE	EA	40.50		
	PRIME AND 1 COAT	EA	66.00		
	PRIME AND 2 COATS	EA	90.00		
	FOR SINGLE DOOR WITH TWO SIDELIGHTS				
	1 COAT ON PAINTED SURFACE	EA	60.00		
	PRIME AND 1 COAT	EA	84.00		
	PRIME AND 2 COATS	EA	108.00		

24. DOOR PAINTING

SPECIFICATIONS		UNIT	PRICE	LOCAL AREA MODIFICATION	
				%	PRICE
GLIDING PATIO DOOR	PAINT ONE SIDE ONLY, INCLUDING EXTERIOR JAMB AND CASINGS, WITH BRUSH, TWO PANELS				
	1 COAT ON PAINTED SURFACE	EA	22.95		
	PRIME AND 1 COAT	EA	38.70		
	PRIME AND 2 COATS	EA	54.45		
	STAIN AND/OR CLEAR FINISH WITH BRUSH				
	1 COAT	EA	29.10		
	2 COATS	EA	49.50		
	3 COATS	EA	69.75		
	SAME AS ABOVE, THREE PANELS WITH BRUSH				
	1 COAT ON PAINTED SURFACE	EA	27.75		
	PRIME AND 1 COAT	EA	46.80		
	PRIME AND 2 COATS	EA	66.45		
	STAIN AND/OR CLEAR FINISH WITH BRUSH				
	1 COAT	EA	35.40		
	2 COATS	EA	59.03		
	3 COATS	EA	85.35		
GARAGE DOOR	BOTH SIDES, INCLUDING JAMBS AND CASINGS, PAINT WITH BRUSH, 9-0 X 7-0				
	1 COAT ON PAINTED SURFACE	EA	82.35		
	PRIME AND 1 COAT	EA	141.60		
	PRIME AND 2 COATS	EA	187.50		
	SAME AS ABOVE, WITH SPRAY GUN				
	1 COAT ON PAINTED SURFACE	EA	36.75		
	PRIME AND 1 COAT	EA	64.50		
	PRIME AND 2 COATS	EA	89.33		
	BOTH SIDES, INCLUDING JAMBS AND CASINGS, PAINT WITH BRUSH, 16-0 X 7-0				
	1 COAT ON PAINTED SURFACE	EA	146.10		
	PRIME AND 1 COAT	EA	250.80		
	PRIME AND 2 COATS	EA	334.50		
	SAME AS ABOVE, WITH SPRAY GUN				
	1 COAT ON PAINTED SURFACE	EA	65.25		
	PRIME AND 1 COAT	EA	114.60		
	PRIME AND 2 COATS	EA	158.70		

SPECIFICATIONS		UNIT	PRICE	LOCAL AREA MODIFICATION	
				%	PRICE
PAINT REMOVAL	REMOVE EXISTING PAINT OR VARNISH WITH CHEMICALS, 1 OR 2 LAYERS PER COAT OF PAINT REMOVER				
	INTERIOR TRIM	LF	1.22		
	PANELING	SF	1.46		
STRIP CABINETS	REMOVE EXISTING PAINT OR VARNISH WITH CHEMICALS				
	BOOKSHELVES OR CABINETS (INCLUDING BACKING)	SF	1.76		
	CABINET DOORS (2 SIDES)	SF	2.55		
	SF = FRONT SQUARE FOOTAGE OF SHELVES, CABINETS OR DOORS				
BURNING OFF	PLAIN SURFACES, NON-LEAD BASED PAINT				
	AVERAGE	SF	1.31		
	EXTENSIVE	SF	1.70		
	INTERIOR TRIM	LF	1.16		
	DE-GLOSS WALLS OR TRIM	SF	.32		
SANDING AND FILLING	FILL HOLES AND SMALL CRACKS IN WALLS OR CEILING				
	AVERAGE	SF	.11		
	EXTENSIVE	SF	.18		
	SAND AND FILL TRIM	LF	.26		
	CAULK TRIM	LF	.59		
HARDWARE	CABINET HARDWARE REMOVAL AND REPLACEMENT	EA	1.50		
	LOCKSET REMOVAL AND REPLACEMENT	EA	10.50		
WASHING	WASH INTERIOR WITH TSP BEFORE REPAINTING				
	SMOOTH WALLS	SF	.17		
	TEXTURED WALLS	SF	.20		
	DOOR JAMB, CASING AND TRIM	EA	7.50		
	WINDOW, INCLUDING ALL TRIM	EA	12.00		
	MOULDINGS	LF	.21		

24. DOOR PAINTING, INTERIOR

SPECIFICATIONS		UNIT	PRICE	LOCAL AREA MODIFICATION	
				%	PRICE
INTERIOR DOOR	PAINT BOTH SIDES OF FLUSH OR PANEL DOOR INCLUDING CASING & JAMB, WITH BRUSH				
	1 COAT ON PAINTED SURFACE	EA	21.00		
	PRIME AND 1 COAT	EA	37.50		
	PRIME AND 2 COATS	EA	54.00		
	WITH ROLLER				
	1 COAT ON PAINTED SURFACE	EA	12.00		
	PRIME AND 1 COAT	EA	21.00		
	PRIME AND 2 COATS	EA	30.00		
	STAIN AND/OR CLEAR FINISH, WITH BRUSH				
	1 COAT	EA	27.00		
	2 COATS	EA	48.00		
	3 COATS	EA	69.00		
LOUVER DOOR	PAINT BOTH SIDES OF LOUVER DOOR, INCLUDING CASING AND JAMB, WITH BRUSH				
	1 COAT ON PAINTED SURFACE	EA	57.00		
	PRIME AND 1 COAT	EA	102.00		
	PRIME AND 2 COATS	EA	147.00		
	STAIN AND/OR CLEAR FINISH, WITH BRUSH				
	1 COAT	EA	73.50		
	2 COATS	EA	131.25		
	3 COATS	EA	189.00		
BIFOLD DOORS	PAINT BOTH SIDES OF BIFOLD UNIT UP TO 5'-0" WIDE OPENING, INCL. CASING AND JAMB, WITH BRUSH				
	1 COAT ON PAINTED SURFACE	EA	46.50		
	PRIME AND 1 COAT	EA	78.00		
	PRIME AND 2 COATS	EA	111.00		
	STAIN AND/OR CLEAR FINISH, WITH BRUSH				
	1 COAT	EA	58.50		
	2 COATS	EA	99.00		
	3 COATS	EA	141.00		

24. WINDOW PAINTING, INTERIOR OR EXTERIOR

SPECIFICATIONS		UNIT	PRICE	LOCAL AREA MODIFICATION	
				%	PRICE
WINDOW	PAINT ONE SIDE ONLY, WOOD OR METAL DOUBLE HUNG, SLIDING, CASEMENT, AWNING OR PICTURE WINDOW, INCLUDING ALL TRIM ONE SIDE, **1-8 TO 3-0 WIDTH**	EA	16.05		
	1 COAT ON PAINTED SURFACE	PLUS LITE	.71		
	PRIME AND 1 COAT	EA	27.60		
		PLUS LITE	1.17		
	PRIME AND 2 COATS	EA	39.75		
	LITE = EACH LITE; COUNT THE LITES AND MULTIPLY BY COSTS	PLUS LITE	1.65		
	3-4 TO 7-0 WIDTH 1 COAT ON PAINTED SURFACE	EA	19.80		
		PLUS LITE	.71		
	PRIME AND 1 COAT	EA	34.50		
		PLUS LITE	1.17		
	PRIME AND 2 COATS	EA	48.00		
		PLUS LITE	1.65		
	STAIN AND/OR CLEAR FINISH, **1-8 TO 3-0 WIDTH** 1 COAT	EA	20.55		
		PLUS LITE	.93		
	2 COATS	EA	34.35		
		PLUS LITE	1.67		
	3 COATS	EA	51.00		
		PLUS LITE	2.30		
	3-4 TO 7-0 WIDTH 1 COAT	EA	33.90		
		PLUS LITE	.93		
	2 COATS	EA	58.50		
		PLUS LITE	1.56		
	3 COATS	EA	82.35		
		PLUS LITE	2.19		
	PER ADDITIONAL STORY, IF LADDER OR SCAFFOLDING IS REQUIRED **ADD**	EA	40%		
	EA = EACH OPENING				
GRILLES	DIVIDED LIGHT GRILLE PAINT OR STAIN, 1 COAT	EA	9.30		
	PAINT OR STAIN, 2 COATS	EA	17.10		

24. INTERIOR PAINTING

SPECIFICATIONS		UNIT	PRICE	LOCAL AREA MODIFICATION	
				%	PRICE
WALL	SMOOTH FINISH PLASTER OR PLASTERBOARD, WITH ROLLER				
	1 COAT ON PAINTED SURFACE	SF	.32		
	PRIME AND 1 COAT	SF	.53		
	PRIME AND 2 COATS	SF	.72		
	SAME AS ABOVE, WITH SPRAY GUN				
	1 COAT ON PAINTED SURFACE	SF	.26		
	PRIME AND 1 COAT	SF	.48		
	PRIME AND 2 COATS	SF	.72		
PANELING	PAINT WOOD PANELLED WALLS, WITH BRUSH				
	1 COAT ON PAINTED SURFACE	SF	.56		
	PRIME AND 1 COAT	SF	.90		
	PRIME AND 2 COATS	SF	1.25		
	PAINT WITH ROLLER				
	1 COAT ON PAINTED SURFACE	SF	.33		
	PRIME AND 1 COAT	SF	.57		
	PRIME AND 2 COATS	SF	.80		
	STAIN AND/OR CLEAR FINISH, ON PANELLED WALLS WITH ROLLER				
	1 COAT	SF	.42		
	2 COATS	SF	.69		
	3 COATS	SF	.98		
MOULDING	BASE, CEILING MOULDING, CHAIR RAIL OR DOOR OR WINDOW TRIM UP TO 6" *(IF SAME COLOR AND FINISH AS WALL OR CEILING, INCLUDE IN WALL OR CEILING ESTIMATE)*				
	1 COAT ON PAINTED SURFACE	LF	.42		
	PRIME AND 1 COAT	LF	.74		
	PRIME AND 2 COATS	LF	1.04		
	SAME AS ABOVE WITH STAIN AND/OR CLEAR FINISH				
	1 COAT	LF	.45		
	2 COATS	LF	.78		
	3 COATS	LF	1.10		
STAIR HANDRAIL	PAINT HANDRAIL, BALUSTERS AND NEWEL POST, ONE FLIGHT				
	1 COAT ON PAINTED SURFACE	EA	66.00		
	PRIME AND 1 COAT	EA	123.00		
	PRIME AND 2 COATS	EA	168.00		

SPECIFICATIONS		UNIT	PRICE	LOCAL AREA MODIFICATION	
				%	PRICE
SMOOTH CEILING	SMOOTH FINISH PLASTER OR PLASTERBOARD, WITH ROLLER				
	1 COAT ON PAINTED SURFACE	SF	.38		
	PRIME AND 1 COAT	SF	.63		
	PRIME AND 2 COATS	SF	.89		
	SAME AS ABOVE, WITH SPRAY GUN				
	1 COAT ON PAINTED SURFACE	SF	.26		
	PRIME AND 1 COAT	SF	.48		
	PRIME AND 2 COATS	SF	.72		
TEXTURED CEILING	TEXTURED CEILING, WITH ROLLER				
	1 COAT ON PAINTED SURFACE	SF	.35		
	PRIME AND 1 COAT	SF	.59		
	PRIME AND 2 COATS	SF	.86		
	SAME AS ABOVE, WITH SPRAY GUN				
	1 COAT ON PAINTED SURFACE	SF	.26		
	PRIME AND 1 COAT	SF	.48		
	PRIME AND 2 COATS	SF	.72		
WOOD CEILING	WOOD TONGUE & GROOVE CEILING, WITH BRUSH				
	1 COAT ON PAINTED SURFACE	SF	.60		
	PRIME AND 1 COAT	SF	.99		
	PRIME AND 2 COATS	SF	1.35		
	SAME AS ABOVE, WITH ROLLER				
	1 COAT ON PAINTED SURFACE	SF	.38		
	PRIME AND 1 COAT	SF	.63		
	PRIME AND 2 COATS	SF	.89		
	SAME AS ABOVE, WITH SPRAY GUN				
	1 COAT ON PAINTED SURFACE	SF	.26		
	PRIME AND 1 COAT	SF	.48		
	PRIME AND 2 COATS	SF	.72		
CATHEDRAL CEILING	SINGLE OR DOUBLE PITCH CEILING				
	Highest Point Above Floor				
	6/12 SLOPE 12'-0" **ADD**	SF	37%		
	16'-0" **ADD**	SF	58%		
	20'-0" **ADD**	SF	90%		
	12/12 SLOPE 12'-0" **ADD**	SF	67%		
	16'-0" **ADD**	SF	85%		
	20'-0" **ADD**	SF	121%		
	SF = PLAN AREA, **NOT** ACTUAL AREA COVERED				

24. INTERIOR PAINTING

SPECIFICATIONS		UNIT	PRICE	LOCAL AREA MODIFICATION %	LOCAL AREA MODIFICATION PRICE
CASED OPENING	DOOR JAMBS AND TWO SIDES CASING, PAINT WITH BRUSH, 3'-0" OPENING				
	1 COAT ON PAINTED SURFACE	EA	20.25		
	PRIME AND 1 COAT	EA	33.75		
	PRIME AND 2 COATS	EA	44.25		
	SAME AS ABOVE, 6'-0" OPENING				
	1 COAT ON PAINTED SURFACE	EA	26.40		
	PRIME AND 1 COAT	EA	45.00		
	PRIME AND 2 COATS	EA	59.10		
	STAIN AND/OR CLEAR FINISH ON 3'-0" OPENING				
	1 COAT	EA	25.50		
	2 COATS	EA	43.50		
	3 COATS	EA	57.00		
	SAME AS ABOVE ON 6'-0" OPENING				
	1 COAT	EA	33.00		
	2 COATS	EA	55.80		
	3 COATS	EA	73.05		
RADIATOR	*NOTE:* TO OBTAIN SQUARE FOOT DIMENSION, MULTIPLY HEIGHT OF RADIATOR BY LENGTH OF RADIATOR, **THE FACE ONLY.** COSTS INCLUDE PAINTING THE ENTIRE RADIATOR WITH BRUSH				
	1 COAT	SF	4.05		
	2 COATS	SF	7.94		
	SAME AS ABOVE, WITH SPRAY GUN				
	1 COAT	SF	1.50		
	2 COATS	SF	2.99		
SHELVES	OPEN SHELVES, **NO** BACKING OR VERTICAL SUPPORTS, TOP, BOTTOM AND EDGES, PAINT WITH BRUSH, UP TO 12" SHELVES				
	1 COAT ON PAINTED SURFACE	LF	.81		
	PRIME AND 1 COAT	LF	1.50		
	PRIME AND 2 COATS	LF	2.21		
	STAIN AND/OR CLEAR FINISH WITH BRUSH				
	1 COAT	LF	.86		
	2 COATS	LF	1.58		
	3 COATS	LF	2.33		

24. INTERIOR SHELVES, TRIM PAINTING

SPECIFICATIONS		UNIT	PRICE	LOCAL AREA MODIFICATION	
				%	PRICE
	TO OBTAIN SQUARE FOOT DIMENSIONS, MULTIPLY HEIGHT OF BOOKCASE, CABINET OR DOOR BY THE WIDTH, **ONE FACE ONLY**. PRICES SHOWN INCLUDE PAINTING ALL SIDES AND EDGES OF SHELVES & TRIM AND ONE SIDE OF BACKING.				
BOOKCASE AND CABINETS	PAINT WITH BRUSH				
	SHELVES 12" DEEP				
	1 COAT ON PAINTED SURFACE	SF	2.61		
	PRIME AND 1 COAT	SF	4.52		
	PRIME AND 2 COATS	SF	5.94		
	SHELVES 24" DEEP		5.22		
	1 COAT ON PAINTED SURFACE	SF			
	PRIME AND 1 COAT	SF	9.03		
	PRIME AND 2 COATS	SF	11.90		
	SAME AS ABOVE WITH STAIN AND/OR CLEAR FINISH				
	SHELVES 12" DEEP				
	1 COAT	SF	3.33		
	2 COATS	SF	5.76		
	3 COATS	SF	7.56		
	SHELVES 24" DEEP				
	1 COAT	SF	6.69		
	2 COATS	SF	11.61		
	3 COATS	SF	15.14		
CABINET DOORS	TWO SIDES AND ALL EDGES OF CABINET DOORS, PAINT WITH BRUSH				
	1 COAT ON PAINTED SURFACE	SF	2.34		
	PRIME AND 1 COAT	SF	4.01		
	PRIME AND 2 COATS	SF	5.43		
	SAME AS ABOVE WITH STAIN AND/OR CLEAR FINISH 1 COAT	SF	2.46		
	2 COATS	SF	4.22		
	3 COATS	SF	5.72		
CLOTHES CLOSET	PAINT SHELF AND TRIM ONLY				
	1 COAT ON PAINTED SURFACE	LF	1.50		
	PRIME AND 1 COAT	LF	2.66		
	PRIME AND 2 COATS	LF	3.90		
	LF = WIDTH OF CLOSET				
LINEN CLOSET	24" DEEP SHELVES, 12" O.C. VERTICALLY MEASURED, PAINT SHELVES & TRIM ONLY				
	1 COAT ON PAINTED SURFACE	LF	22.50		
	PRIME AND 1 COAT	LF	38.85		
	PRIME AND 2 COATS	LF	55.50		
	LF = WIDTH OF CLOSET				

24. WALLPAPERING

SPECIFICATIONS		UNIT	PRICE	LOCAL AREA MODIFICATION	
				%	PRICE
WALLPAPER	WALLPAPER APPLIED TO WALLS AND/OR CEILING, CEILING UP TO 8 FEET ABOVE FLOOR				
	Retail Cost of Wallpaper				
	$11 PER ROLL	SF	1.26		
	$14 PER ROLL	SF	1.55		
	$17 PER ROLL	SF	1.85		
	$20 PER ROLL	SF	2.13		
	$25 PER ROLL	SF	2.55		
	$30 PER ROLL	SF	2.96		
VINYL	SAME AS ABOVE, WITH VINYL WALL COVERING				
	LIGHT	SF	1.35		
	MEDIUM	SF	1.68		
	HEAVY	SF	2.06		
GRASS CLOTH	SAME AS ABOVE, WITH GRASS CLOTH				
	ECONOMY	SF	2.10		
	MEDIUM	SF	2.61		
	PREMIUM	SF	3.63		
HIGH CEILING	FOR HANGING PAPER IN ROOMS WITH CEILING MORE THAN 8 FEET ABOVE FLOOR (UP TO 12 FEET), ADD TO ABOVE COSTS **ADD**	SF	10%		
OVER OLD PAPER	FOR HANGING PAPER OVER OLD LAYER(S) OF WALLPAPER **ADD**	SF	10%		
DIFFICULT JOB	FOR DIFFICULT JOB WITH MORE THAN AVERAGE AMOUNT OF CUTTING **ADD**	SF	10%		
SMALL JOB	FOR HANGING PAPER IN SMALL ROOF WITH LESS THAN 200 SF OF WALL AREA **ADD**	SF	5%		
REMOVE WALLPAPER	REMOVE WALLPAPER FROM PLASTER OR DRY-WALL WITH STEAMING EQUIPMENT				
	ONE LAYER	SF	.35		
	SEVERAL LAYERS	SF	.54		
	WASH OFF GLUE AFTER REMOVING PAPER FROM WALL	SF	.20		

SPECIFICATIONS		UNIT	PRICE	LOCAL AREA MODIFICATION %	LOCAL AREA MODIFICATION PRICE
	NOTE: PRICES SHOWN BELOW INCLUDE REMOVAL OF RUBBISH FROM PREMISES TO DUMPING GROUND, BUT **NO** TEAR-OUT COSTS AND **NO** DUMPSTER OR DUMPING FEES				
ADDITION	INCLUDING AVG. AMOUNT OF TEAR-OUT DEBRIS	EA PLUS	186.00		
	SF = ROOM ADDITION	SF	.39		
BATHROOM AND KITCHEN	INCLUDING REMOVAL OF OLD TILE, PLUMBING FIXTURES, CABINETS AND APPLIANCES (DOES **NOT** INCLUDE DISCONNECTING FIXTURES)	EA	123.00		
	EA = ONE LOAD OF RUBBISH (5 YARDS)				
DORMERS	INCLUDES REMOVAL OF OLD ROOFING AND FRAMING MATERIALS FROM PREMISES	EA PLUS	372.00		
		SF	.39		
	SF = SIZE OF NEW DORMER				
GARAGE OR CARPORT	REMOVAL OF SCRAP LUMBER, MASONRY AND TEAR-OUT DEBRIS	EA	186.00		
	EA = EACH JOB				
PORCH	REMOVAL OF SCRAP LUMBER AND MASONRY	EA	105.00		
BASEMENT FAMILY ROOM	REMOVAL OF SCRAP LUMBER & TEAR-OUT DEBRIS	EA PLUS	105.00		
	SF = FAMILY ROOM	SF	.36		
PORCH OR CARPORT ENCLOSURE	REMOVAL OF SCREENING, POSTS, AND OTHER TEAR-OUT DEBRIS	EA	159.00		

25. CLEAN-UP AND HAULING

SPECIFICATIONS		UNIT	PRICE	LOCAL AREA MODIFICATION	
				%	PRICE
DEMOLITION	REMOVAL OF DEBRIS FROM DEMOLITION WORK, LOADING MASONRY, PLASTER, LUMBER AND OTHER TEAR-OUT DEBRIS -- LOAD TRUCK FROM BUILDING BY HAND AND HAUL TO DUMPING GROUND WITHIN 5 MILES DUMPING FEES AND DUMPSTER COST **NOT** INCLUDED SF = FLOOR AREA OF BUILDING	EA PLUS SF	546.00 .54		
	SAME AS ABOVE, CHUTE RUBBISH FROM BUILDING AND HAUL	EA PLUS SF	546.00 .36		
DIRT	SHOVEL DIRT BY HAND INTO TRUCK FROM PILES AND HAUL TO DUMPING GROUND UP TO 5 MILES EA = 5 CU. YARDS OF DIRT (ONE TRUCKLOAD)	EA	267.00		
ROOFING	FOR REMOVAL OF ROOF COVERING FROM ROOF, SEE PAGE 124				
	REMOVAL OF SCRAP LUMBER AND HAULING UP TO 5 MILES	EA PLUS SF	84.00 .18		
MAID SERVICE	COMPLETE CLEANING OF INTERIOR OF AREA REMODELED, INCLUDING WINDOWS INSIDE AND OUT	EA PLUS SF	75.00 .27		

25

INDEX

INDEX

INDEX

INDEX

INDEX

INDEX

INDEX